ERYTHROPOIETIN AND THE NERVOUS SYSTEM

Novel Therapeutic Options for Neuroprotection

ERYTHROPOIETIN AND THE NERVOUS SYSTEM

Novel Therapeutic Options for Neuroprotection

Edited by Ahmet Höke M.D., Ph.D.

Associate Professor of Neurology and Neuroscience
Johns Hopkins University, School of Medicine

Springer

Cover illustration: Dorsal root ganglion sensory neurons stained with anti-βIII tubulin antibody (green) over Schwann cells stained with anti-erythropoietin antibody (red).

Library of Congress Control Number: 2005934912

ISBN-10: 0-387-30010-4 e-ISBN-10: 0-387-30011-2
ISBN-13: 978-0-387-30010-8

Printed on acid-free paper.

Printed in the United States of America.

9 8 7 6 5 4 3 2 1

springeronline.com

Dedicated to my parents, Gülsen and Safa Höke

LIST OF CONTENTS

List of Contributors

Mathias Bähr MD: Center for Neurological Medicine, University of Göttingen, Robert-Koch-Str. 40, 37075 Göttingen, Germany mbaehr@gwdg.de

Christian Bauer MD: Physiologisches Institut, Universität Zurich-Irchel, Winterhurerstrasse 190, CH-8057 Zurich, Switzerland. cbauer@access.unizh.ch

Paolo Bigini PhD: Lab. Neuroimmunology, "Mario Negri" Institute, via Eritrea 62, 20157 Milano, Italy bigini@marionegri.it

Michael Brines MD, PhD: The Kenneth S. Warren Institute, 712 Kitchawan Road, Ossining, New York 10562, USA. mbrines@kswi.org

Wendy Campana PhD: University of California San Diego, Department of Anesthesiology,9500 Gilman Dr., La Jolla, California 92093-0629, wcampana@ucsd.edu

Anthony Cerami PhD: The Kenneth S. Warren Institute, 712 Kitchawan Road, Ossining, New York 10562, USA. acerami@kswi.org

Juan C. Chavez PhD: Burke/Cornell Medical Research Institute, 785 Mamaroneck Ave. White Plains, NY 10605 USA. jchavez@burke.org

Eric J. Demers MD: University of Washington, Department of Pediatrics, Box 356320, 1959 NE Pacific St., Seattle, Washington 98195-6320, USA. edemers@u.washington.edu

Murat Digicaylioglu MD, PhD: University of California San Diego, Department of Neurosurgery 200 West Arbor Drive-8893, San Diego, CA 92103, USA. muratd@mindspring.com

Hannelore Ehrenreich MD: Max-Planck-Institut für experimentelle Medizin, Hermann-Rein Str.3, 37075 Göttingen, Germany ehrenreich@em.mpg.de

Francis X. Farrell PhD: Johnson and Johnson P.R.D. L.L.C., Centocor Discovery, Radnor, Pennsylvania 19087, USA mffarrell@cntus.jnj.com

JoAnn M. Gensert PhD: Burke/Cornell Medical Research Institute, 785 Mamaroneck Ave. White Plains, NY 10605 USA. jgensert@burke.org

Pietro Ghezzi PhD: Head Lab. Neuroimmunology, "Mario Negri" Institute, via Eritrea 62, 20157 Milano, Italy. ghezzi@marionegri.it

Ahmet Höke MD, PhD: Johns Hopkins University, Department of Neurology, 600 N. Wolfe St. Path 509, Baltimore, Maryland 21287, USA. ahoke@jhmi.edu

Linda K. Jolliffe PhD: Johnson and Johnson P.R.D. L.L.C., Drug Discovery Section, Raritan, New Jersey 08869, USA ljolliff@prdus.jnj.com

Sandra Juul MD, PhD: University of Washington, Department of Pediatrics, Box 356320, 1959 NE Pacific St., Seattle, Washington 98195-6320, USA. sjuul@u.washington.edu

Sanjay C. Keswani MBBS: Johns Hopkins University, Department of Neurology, 600 N. Wolfe St. Path 627, Baltimore, Maryland 21287, USA. skeswani@jhmi.edu

Marcel Leist PhD: Department of Disease Biology, H. Lundbeck A/S, 9 Ottiliavej, DK-2500 Valby, Copenhagen, Denmark male@lundbeck.com

Hugo H. Marti MD: University of Heidelberg, Institute of Physiology and Pathophysiology, Im Neuenheimer Feld 326, D-69120 Heidelberg, Germany hugo.marti@physiologie.uni-hd.de

Giorgia Melli MD: Johns Hopkins University, Department of Neurology, 600 N. Wolfe St. Path 509, Baltimore, Maryland 21287, USA. gmelli2@jhmi.edu

Manuela Mengozzi PhD: Lab. Neuroimmunology, "Mario Negri" Institute, via Eritrea 62, 20157 Milano, Italy. mengozzi@marionegri.it

Michael J. Renzi PhD: Johnson and Johnson P.R.D. L.L.C., Neurological Disorders Section, Raritan, New Jersey 08869, USA mrenzi@prdus.jnj.com

Kenneth J. Rhodes PhD: Johnson and Johnson P.R.D. L.L.C., Neurological Disorders Section, Raritan, New Jersey 08869, USA krhodes2@prdus.jnj.com

Gundula Rohde: Center for Neurological Medicine, University of Göttingen, Robert-Koch-Str. 40, 37075 Göttingen, Germany grohde@gwdg.de

Anna-Leena Siren MD: Max-Planck-Institut für experimentelle Medizin, Hermann-Rein Str.3, 37075 Göttingen, Germany siren@em.mpg.de

Lars Torup PhD: Department of Neuropharmacology, H. Lundbeck A/S, 9 Ottiliavej, DK-2500 Valby, Copenhagen, Denmark lto@lundbeck.com

Jochen H. Weishaupt MD: Center for Neurological Medicine, University of Göttingen, Robert-Koch-Str. 40, 37075 Göttingen, Germany jweisha@gwdg.de

PREFACE

The idea for this book originated at the annual Society for Neuroscience meeting in the fall of 2003. As someone new to the field, I was pleasantly surprised to see many investigators from diverse disciplines working on erythropoietin. Even though erythropoietin was initially identified as the growth factor that induced red blood cell proliferation, it has multiple actions on a diverse population of cells and tissues, as is the case with many growth factors. What sets erythropoietin apart is the fact that it is one of the earliest recombinant proteins that has been in clinical use as a drug. This past experience with dosing and side effect profile makes it an ideal candidate for further development for neuroprotective therapies. In this book we have strived to bring a current state-of-the-art review of multiple aspects of erythropoietin research as it relates to the nervous system. Our hope is that this book will stimulate new research on erythropoietin and the nervous system and bring new investigators to the field.

First, I would like to thank all of the contributors to the book. Without their efforts and patience we could not have finished this book. I would also like to thank Marcia Kidston at Springer US and my assistant, Stephanie David for help and patience in getting the chapters formatted properly. Finally many thanks go to my wife, Nishi and my children, Maya and Erol, for their unwavering support and love.

Ahmet Höke MD, PhD
Baltimore, 2005

Chapter 1

HISTORY AND BIOLOGY OF ERYTHROPOIETIN IN HEMATOPOIETIC AND NON-NEURAL TISSUES

Giorgia Melli, Sanjay C. Keswani and Ahmet Höke
Department of Neurology, Johns Hopkins University School of Medicine, Baltimore, MD, USA

Abstract: Erythropoietin (EPO) is a glycoprotein hormone, which is produced in kidney and liver, and is mainly involved in regulating proliferation and maturation of red blood cells. EPO gene expression is induced by hypoxia through the transcription factor hypoxia inducible factor-1, which has been found to be the main regulator of oxygen homeostasis in the body. Suppression of apoptosis is the principle mechanism of action of EPO in maintaining erythropoiesis. It has been recently recognized that EPO is a member of the cytokine type I superfamily and it has multiple effects in organs and tissues different from the hematopoietic system. Recent evidence of EPO as a protective factor in various injury models in the nervous system and heart has raised the possibility that EPO can exert protective effects in many organs in the body. However, whether the mechanism of protective action involves inhibition of apoptosis remains to be seen.

Key words: erythropoietin, history, erythropoiesis, hypoxia inducible factor-1, apoptosis, and tissue-protection

1. INTRODUCTION

The hormone EPO is an essential growth factor for the production of red blood cells. In healthy adult humans, one percent of all red blood cells are destroyed every day and replaced by reticulocytes. Blood oxygen availability is the principle regulator of erythropoiesis; hypoxia induces EPO gene expression in the kidneys and liver (Jelkmann et al., 2001). The plasma level of EPO may rise 1000-fold above the normal value and as a consequence the basal rate of production of red cells (2-3 x 10^{11} per day) may increase 10-fold in hypoxic stress (Jelkmann, 1992).

In the recent years EPO and its receptor (EPOR) have been shown to be present in other tissues not involved in hematopoiesis. These include the brain, the reproductive tract, (Marti et al 1996, Kobayashi et al 2002, Masuda et al 2000), the lung, the spleen and the heart (Fandrey et al, 1993). These discoveries raised the possibility of EPO acting not only as an erythropoietic hormone but also as a possible protective factor in many organs. In particular, increased expression of EPO and EPOR in the brain following an ischemic injury (Siren et al, 2001), suggests a more likely paracrine rather than an endocrine mode of action of EPO at these sites. Moreover EPO in the brain and testis may be separated from the systemic circulation by the blood-brain or blood-testis barrier, even if it has been shown that EPO can cross the blood-brain barrier especially after brain damage (Fandrey, 2004).

2. HISTORICAL NOTES

In 1882 Paul Bert, a pupil of Claude Bernard's school in Paris, described for the first time an increased number of red blood cells and consequently an increased blood O_2 capacity in animals living at high altitudes and considered it as genetically derived (Bert, 1882). He was also the one who showed that the mountain sickness is due to hypoxemia (Bert, 1878). Few years later in 1890, the French histologist Viault showed that red blood cell production was inducible by permanence at altitudes; in particular he noted a significant increase in erythrocytes in his blood during a 3 weeklong expedition to the Peruvian mountains (Viault, 1890). On the basis of these early studies, Friedrich Miescher suggested that low oxygen pressure acted directly on the bone marrow to stimulate the production of red blood cells (Miescher, 1893).

In 1906 Carnot and DeFlandre published an intriguing work where they postulated that a humoral factor, which they called *hémopoïétin*, regulated red blood cell production (Carnot et al, 1906) They carried out experiments which showed a prompt increase in reticulocytes in normal rabbits following the injection of plasma from donor rabbits who have had a bleeding stimulus. In the following years other investigators failed to confirm the results of Carnot and DeFlandre and the existence of a hemopoietic hormone was not generally accepted. However, in 1948 two Finnish scientists, Bonsdorff and Jalavisto continued work on red blood cell production and called the hemopoietic substance "erythropoietin" (Bonsdorff E et al, 1948). The most important works on the existence of EPO were the classical experiments of Reissman (Reissmann, 1950) and Erslev (Erslev, 1953) on parabiotic rats. Reissmann showed that erythroid hyperplasia in bone marrow and reticulocytosis were induced in both parabiotic animals when only one partner was exposed to hypoxemia: this was an elegant demonstration that a circulating substance was able to stimulate red blood cell production in an animal exposed to a normal atmospheric pressure. Finally, Erslev confirmed the early experiment of Carnot and DeFlandre, by inducing reticulocytosis, and later increased hematocrit, in rabbits repeatedly injected with large volumes of plasma from anemic donor animals. In his work Erslev also predicted the potential therapeutic role of EPO in treating anemia. The following step was the important and laborious work of purification of human EPO from urine of anemic patients by Miyake in 1977 (Miyake et al, 1977). It allowed the successful cloning and transfection in mammalian cells of the EPO gene in 1985 by Lin in Chinese hamster ovary cells (Lin et al, 1985) and by Jacobs in African green monkey kidney cells (Jacobs et al, 1985). It permitted the consequent industrial production of recombinant human erythropoietin for treating patients with anemia.

The questions about the site(s) of production of EPO started to be answered when Jacobson reported that bilaterally nephrectomized animals, subjected to bleeding, failed to produce increased EPO (Jacobson et al, 1957). The same observation was also reported in anemic patients affected by chronic renal failure (Gurney et al, 1957). In 1961, Fisher demonstrated that cobalt enhanced production of EPO in the isolated perfused dog kidney (Fisher et al, 1961). Few months later Kuratowska published an experiment in rabbits showing that the isolated kidney perfused with hypoxemic blood increased EPO production, using a reticulocyte assay for the assessment of EPO activity (Kuratowska et al, 1961). However for many years the issue of kidneys as site of synthesis of EPO was still debated. It was thought that kidney under hypoxic stimulus released an enzyme, "erythrogenin", which was able to activate a hepatic precursor of erythropoietin (Gordon et al 1967). This theory was finally abandoned when several different experiments converged to show that EPO was synthesized in kidneys: erythropoietin activity was quantified in the serum-free medium of isolated perfused kidneys in rabbit (Erslev, 1974) and in renal extracts of rats exposed to hypoxia (Jelkmann et al, 1981). The final striking evidence of renal production of EPO was given by the demonstration of EPO mRNA in renal extracts (Beru et al, 1986; Schuster et al, 1987).

In contrast to the adults, liver is the principal site of production of EPO during fetal life (Zanjani et al, 1974). Even though the kidneys take over as the main site of production after birth, liver continues to contribute to the synthesis of EPO in a minor role (Fried, 1972).

3. STRUCTURE AND MECHANISM OF ACTION OF EPO IN ERYTHROPOIESIS

EPO is a glycoprotein composed of 165 aminoacids and 4 carbohydrate side chains with an estimated molecular mass of 34 kD; 40% of which is carbohydrate (Jekjmann, 2003). There are 3 tetraantennary N-linked and 1 small O-linked acid sugar chains. The molecule forms a bundle of 4 α-helices, which are folded into a compact globular structure (Jelkmann, 2003). The carbohydrate moiety, rich in sialic acid, is critical to molecular stability and full in vivo biological activity. In fact, the form of EPO deprived of sialic acid is rapidly sequestered in the liver (Spivak et al, 1989).

The activity of EPO starts by binding to its receptor EPOR. In the absence of EPOR, EPO does not have any erythropoietic activity. Transgenic mice lacking EPOR develop severe anemia (Kieran et al 1996, Lin et al 1996). Human EPOR has been cloned; it is a 484 amino acid glycoprotein and a member of the type I superfamily of single-transmembrane cytokine-receptors (Wojchowski et al, 1999). EPOR is expressed in a variety of cells and organs including placenta, endothelium, and megakaryocytes but the primary target cells are the erythroid progenitors, Burst–forming Unit (BFU-E) and Colony-forming Unit (CFU-E). These erythroid progenitors are derived from the stochastic differentiation of bipotential or multipotential progenitors, a population of stem cells (Suda et al, 1984). The CFU-E cells are the most sensitive cells responding to EPO. Although the EPOR is expressed in both populations of erythroid precursors, the number is highest in CFU-E cells, where it has a density over 1000 receptors per cell. The number of receptors per cell gradually declines in more mature cells (Broudy et al, 1991). The primary mechanism of action of EPO in regulating erythropoiesis is the suppression of apoptosis; as a consequence, the progenitor cells proliferate and differentiate, resulting in an increased formation of normoblasts and finally reticulocytes (Jelkmann, 2003). It has been shown that CFU-E cells do not survive in vitro in absence of EPO. Since the majority of CFU-E cells are cycling, their survival in the presence of EPO may be tightly linked to their proliferation and differentiation to mature erythrocytes (Sieff et al, 1986). EPO also is necessary for the survival and terminal differentiation of a subset of BFU-E cells, while a second subset of BFU-E cells, presumably less mature, survive in absence of EPO if other growth factors like IL-3 or GM-CSF are present (Sieff et al, 1989). In addition to stimulating the proliferation and differentiation of erythroid progenitors, EPO activates the mitotic division of proerythroblasts and basophilic erythroblasts and it accelerates the release of reticulocytes from the bone marrow (Jelkmann,

1992). After an acute increase of plasmatic EPO it takes generally 3-4 days to detect an increase in reticulocytes (Jelkmann et al, 2001).

After binding of EPO to its receptor, a tighter homodimerization of the receptors ensue and a conformational change of the intracellular domains activates receptor associated Janus kinase 2 (JAK2) by reciprocal tyrosine phosphorylation. Activated JAK2 phosphorylate the EPOR and several distinct intracellular signaling molecules. The most important down-stream protein transmitting EPO signals is STAT5 (Signal Transducer and Activator of Transcription). STATs are latent cytoplasmic transcription factors, as soon as they are activated by JAKs, they dimerize and translocate into the nucleus where they bind to specific DNA sequences and allow the transcription of the respective genes (Jelkmann et al, 2001). Suppression of apoptosis is the main mechanism of action of EPO in erythropoiesis and there is evidence in favor of a role of STAT5 in the anti-apoptotic pathway of EPO. In particular, STAT5 appears to mediate the induction of Bcl-xL, an antiapoptotic gene through direct binding to its promoter (Silva et al, 1999). In fact, anemia and high levels of apoptosis in erythroid progenitors are seen in STAT5 knockout mice (Socolovsky et al, 1999). Homodimerization and activation of anti-apoptotic signaling via JAK2 and STAT5 are shared by other receptors, including those for thrombopoietin, granulocyte colony-stimulating factor, prolactin and growth hormone. Receptors for each of these growth factors has been reported to homodimerize, bind JAK2 and activate STAT5 (Wojchowski et al, 1999).

In addition to the JAK2-STAT5 pathway, other intracellular pathways are activated by EPO. These include activation of voltage-sensitive calcium channels via EPOR; by this mechanism EPO might have a role in modulation of neurotransmitters (Kawakami et al, 2001). Another pathway involves activation of phosphatidylinositol-3 kinase (PI3-K) and Akt (Kashii et al, 2001), which in turn lead to upregulation of Bcl-xL and inhibition of apoptosis in Baf-3 cells (Leverrier et al, 1999) as well as activation of NF-kB, which mediates a variety of anti-apoptotic signaling pathways. These pathways, along with the JAK2/STAT5 pathway may play an important role in neuroprotection (further details are elsewhere in the book and also reviewed in Ghezzi et al, 2004).

In summary, prevention of apoptosis in late erythroid progenitors is the principle mechanism of action of EPO in erythropoiesis. Although EPO can support the proliferation of murine erythroid cells (Miller et al, 1999) and induce the entry of erythroid progenitors into cell cycle if dormant (Spivak et al, 1991), the underlying molecular events are only partially understood (Jelkmann et al, 2001).

4. REGULATION OF EPO SYNTHESIS

Tissue hypoxia is the primary stimulus for the production of EPO. The cells producing EPO appear to respond to changes in the oxygen capacity, tension and affinity of the blood (Jelkmann, 1992). Hypoxia induces expression of the gene encoding for EPO and research in this field has become the prototype of oxygen regulated gene expression. Most of the current knowledge has been derived from in vitro studies utilizing EPO producing human hepatoma cell cultures and it is possible that the mechanism by which human hepatoma cells regulate EPO production differs from the O2 sensing mechanism and control of EPO production in the kidney. In fact, EPO gene expression in liver occurs in a graded fashion, while in the kidney it follows an all or none rule; the increase in EPO mRNA with hypoxia is due to recruitment of additional cells, all of which are maximally active (Kouri et al, 1989). Crucial in the hypoxia-signaling cascade is the transcription factor complex, hypoxia inducible factor-1 (HIF-1). HIF-1 binds to the hypoxia-responsive element (HRE) in the 3'-flanking enhancer of the EPO gene. It is now known that HIF-1 controls more than 50 oxygen dependent genes and it is the major regulator of oxygen homeostasis in the body. It is a protein heterodimer consisting of the two basic-helix-loop-helix proteins, 120 kD HIF-1 α and 91-94 kD HIF-1 β (Wang et al, 1995). The HIF-1 α subunit is present in the cytoplasm and is unstable at high pO_2; while the HIF-1β subunit is more stable and is permanently present in nuclei. The HIF-1 α subunit has two oxygen dependent degradation domains that target the protein for an immediate ubiquitination and proteosomal degradation under normoxemia (Salceda et al, 1997). Binding of the tumor suppressor protein von-Hippel-Lindau (VHL) to the two oxygen dependent degradation domains of HIF-1 α has a critical role in the degradation process (Maxwell et al, 1999). Enzymatic hydroxylation of HIF-1 α at proline residues is necessary for recognition by VHL (Ivan et al, 2001). The mammalian HIF prolyl hydroxylases have been cloned and designated "prolyl hydroxylase domain" (PHD) containing 1, 2 and 3 (Epstein et al, 2001). The activity of PHDs depends on the availability of oxygen and it is likely that these enzymes are crucial in oxygen sensing regulation of EPO gene transcription (Fandrey, 2004). The observations that cobalt ions and iron chelators such as desferoxamine prevent the formation of the VHL complex (Maxwell et al, 1999) may explain why those compounds stimulate the synthesis of EPO. While HIF-1α is undetectable in cells during normal pO2, the HIF-1α/ β complex can be demonstrated in nuclei within minutes after the induction of hypoxia in cells (Jelkmann et al, 2001).

In addition to erythropoiesis, HIF-1 regulates genes for neoangiogenesis (for example VEGF, vascular endothelial growth factor) and vascular tone (for example nitric oxide synthases). Erythropoiesis and angiogenesis represent adaptive responses to hypoxia that require several days to develop, but HIF-1 also regulates short-term responses to hypoxia, such as the induction of glucose transporters and glycolytic enzymes. Consequently it is possible to argue that HIF-1 plays a role in the metabolic adaptation to hypoxemia represented by the switch of ATP generation from oxidative phosphorylation to glycolysis (Ghezzi et al, 2004). HIF-1 binding is also induced by insulin and insulin-like growth factor 1 and 2, by interleukin 1 β (IL-1 β), and tumor necrosis factor a (TNF-α); thus it is possible to conclude that there is a synergy in the cellular responses to hypoxia, glucose deficiency and inflammation (Jelkmann et al, 2001). However, the interplay between hypoxia, inflammation and glucose deficiency in regulating EPO production are complex. The ability of IGF-1/2, IL-1 β or TNF-α to increase HIF-1 α / β - DNA binding does not necessarily mean that these stimuli result in increased EPO gene expression automatically. In fact, it has been shown that pro-inflammatory cytokines IL-1 and TNF- α lower EPO production (Fandrey et al, 1994; Faquin et al, 1992) while they upregulate other HIF-1 dependent genes like those encoding VEGF or iNOS (inducible NO-synthase) (Hellwig-Burgel et al, 1999). The mechanism by which EPO transcription is reduced during inflammation, despite the increase of HIF-1 binding is not completely understood, but is likely to be due to lowered HNF-4 (hepatocyte nuclear factor 4), a positive transcription factor), and elevated levels of negative regulating transcription factors, GATA-2 and NF-κB (Jelkmann et al 2001; Fandrey, 2004).

H_2O_2 and other reactive oxygen species are responsible for inhibiting EPO gene expression at high pO_2. It has been proposed that H_2O_2 lower EPO expression by activation of GATA-2 (Imagawa et al, 1996). Binding to the GATA-motif, located in the region relative to the transcriptional initiation site of EPO promoter, GATA-2 inhibits the transcription of the gene and it is one of the main candidates to cause the repression of EPO gene under normal levels of pO_2 (Jelkmann et al, 2001).

5. SITES OF SYNTHESIS: KIDNEY AND LIVER

During the fetal life, liver is the principle producer of EPO, but in the adult, expression of EPO is mainly localized to the kidneys. The mechanism and factors determining the switch and the degree of contribution of the liver to the total body production of EPO are different in different species and not completely understood (Fandrey, 2004). Recently it has been proposed that

GATA-4 may be one of the factors that specifically promote EPO expression in the liver (Dame et al, 2004). EPO expression is present in all lobes of the human liver and the cells producing EPO present an expression pattern that follows the pO_2 distribution: high expression around central veins at lower pO_2 and expression extending to the periportal field in anemia or hypoxia (Fandrey, 2004). Since the hepatic expression is so strictly regulated by the pO_2 gradient, it was thought that changes of this gradient around birth could control the switch of production from liver to kidney. But several experiments including surgical procedures aimed at reducing hepatic or renal oxygenation have shown that it does not affect the switch. Thus it is more likely that tissue factors increasing in kidney during and after birth induce EPO expression while repressive factors are expressed in the liver of the adult (Fandrey 2004).

The EPO producing cells in the kidney have been identified in fibroblast-like type I interstitial cells located in the peritubular space in the cortex and outer medulla (Bachmann et al, 1993; Chandel et al, 1998). However until now the attempts to isolate renal EPO producing cells and to set-up renal cell lines for studying the regulation in expression of EPO gene have failed, thus most of the current knowledge derives from in vitro studies on human hepatoma cell cultures.

6. EFFECTS OF EPO IN ORGANS OTHER THAN HEMATOPOIETIC TISSUES

In recent years, there has been increasing evidence that EPO has multiple effects on many tissues and organs far beyond the regulation of erythropoiesis. In the embryo, EPO is a major regulator of vascular formation and organ growth and EPOR have been found in almost every embryonic tissue. EPO and EPOR are also present in many adult tissues. Based on these observations, the concept of a paracrine and/or autocrine action of EPO has been suggested (Bahlmann et al, 2004).

EPO expression has been found in the brain, reproductive tract, lung, spleen, heart and bone marrow. Of particular relevance for the clinical implications is the observation that EPO has a direct activity on endothelial cells. Studies in vitro have shown that EPO increases endothelial cell proliferation and protects endothelial cells against apoptosis (Anagnostou et al, 1990; Carlini et al, 1999). Also human endothelial lines express EPOR and differentiate into vascular structures when exposed to EPO (Carlini et al, 1995; Ribatti et al, 1999). By stimulating mitogenesis and supporting angiogenesis, EPO can improve tissue oxygenation. Recently bone marrow derived endothelial progenitors cells (EPCs), which promote vascular regeneration, have been isolated from the blood (Shi et al, 1998). EPCs are considered to originate from CD34+ stem cells, which can differentiate into erythrocytes, thrombocytes, leukocytes and endothelial cells (Bahlmann et al, 2004). Recently, it has been shown that EPO stimulates angiogenesis, partly by increasing mobilization of EPCs from the bone marrow (Heeschen et al, 2003). In patients with myocardial infarction the clinical outcome correlates with the number of circulating EPCs (Vasa et al, 2001). Moreover it seems that even in subjects without clinically evident cardiovascular disease the number of EPCs correlates with endothelial functions and cardiovascular risk factors (Hill et al, 2001).

EPOR has been demonstrated to be present in the embryonic and fetal heart (Juul et al, 1998) and EPO seems to be involved in cardiac morphogenesis (Wu et al, 1999), even if expression of EPO in the heart appears to be low (Fandray, 2004). Some authors have recently demonstrated that EPO prevents hypoxia-induced apoptosis in neonatal rat ventricular myocytes and they have suggested an Akt dependent pathway (Tramontano et al, 2003) Another study involving rodent isolated hearts have shown that intermittent hypoxia or EPO administration protect against post-ischemic injury (Cai et al, 2003). In this study, during intermittent hypoxia, there was an increase in EPOR mRNA in the heart, but EPO mRNA was at the limit of detection. As predicted, renal EPO mRNA expression and plasma EPO levels were significantly induced by hypoxia, suggesting a probable endocrine mechanism of cardiac protection (Cai et al, 2003). This is in contrast to the nervous system, where astrocytes produce EPO and neurons express EPOR in response to hypoxia (Bergeron et al, 2000; Morishita et al, 1997), suggesting a paracrine mechanism of protection (further discussion of neuroprotection in ischemia is elsewhere in the book).

Other sites of tissue protection by EPO involve the skin and reproductive organs. Recently, experiments in rat skin using flap models have raised the possibility of tissue protection from ischemia by EPO (Buemi et al, 2002; Saray et al, 2003). Finally, there is evidence of EPO expression in human endometrium throughout the menstrual cycle, with higher levels in the secretory than proliferative phase (Yokomizo et al, 2002). Functionally EPO appears to play an important role for endometrial angiogenesis and consequently reproductive function (Fandray, 2004).

REFERENCES

Anagnostou A, Lee ES, Kessimian N, Levinson R, Steiner M. Erythropoietin has a mitogenic and positive chemotactic effect on endothelial cells. Proc Natl Acad Sci USA 1990;87:5978 –5982

Bachmann S, Le Hir M, Eckardt KU. Co-localization of erythropoietin mRNA and ecto-5'-nucleotidase immunoreactivity in peritubular cells of rat renal cortex indicates that fibroblasts produce erythropoietin. J Histochem Cytochem. 1993 Mar;41(3):335-41.

Bahlmann FH, de Groot K, Haller H, Fliser D. Erythropoietin: is it more than correcting anaemia? Nephrol Dial Transplant. 2004 Jan;19(1):20-2.

Bergeron M, Gidday JM, Yu AY, Semenza GL, Ferriero DM, Sharp FR. Role of hypoxia-inducible factor-1 in hypoxia-induced ischemic tolerance in neonatal rat brain. Ann Neurol. 2000 Sep;48(3):285-96.

Bert P. La pression baromètrique. Recherches de physiologie expérimentale. Paris : Masson. 1878

Bert P. Sur la richesse en hèmoglobine du sang des animaux vivant sur le hauts lieux. C R Acad Sci Paris. 1882; 94 : 805-807

Beru N, McDonald J, Lacombe C, Goldwasser E. Expression of the erythropoietin gene. Mol Cell Biol. 1986 Jul;6(7):2571-5.

Bonsdorff E and Jalavisto E. A humoral mechanism in anoxic erythrocytosis. Acta Physiol Scand. 1948; 16: 150-170

Broudy VC, Lin N, Brice M, Nakamoto B, Papayannopoulou T. Erythropoietin receptor characteristics on primary human erythroid cells. Blood. 1991 Jun 15;77(12):2583-90.

Buemi M, Vaccaro M, Sturiale A, Galeano MR, Sansotta C, Cavallari V, Floccari F, D'Amico D, Torre V, Calapai G, Frisina N, Guarneri F, Vermiglio G. Recombinant human erythropoietin influences revascularization and healing in a rat model of random ischaemic flaps. Acta Derm Venereol. 2002;82(6):411-7.

Cai Z, Manalo DJ, Wei G, Rodriguez ER, Fox-Talbot K, Lu H, Zweier JL, Semenza GL. Hearts from rodents exposed to intermittent hypoxia or erythropoietin are protected against ischemia-reperfusion injury. Circulation. 2003 Jul 8;108(1):79-85.

Carlini RG, Alonzo EJ, Dominguez J, Blanca I, Weisinger JR, Rothstein M, Bellorin-Font E. Effect of recombinant human erythropoietin on endothelial cell apoptosis. Kidney Int 1999;55:546 –553

Carlini RG, Reyes AA, Rothstein M. Recombinant human erythropoietin stimulates angiogenesis in vitro Kidney Int 1995; 47:740 –745

Carnot P and Deflandre C. Sur l'activité hémopoiétique du serum au cours de la régénération du sang. C R Acad Sci Paris 1906 ; 143 : 384-386

Chandel NS, Maltepe E, Goldwasser E, Mathieu CE, Simon MC, Schumacker PT. Mitochondrial reactive oxygen species trigger hypoxia-induced transcription. Proc Natl Acad Sci U S A. 1998 Sep 29;95(20):11715-20.

Dame C, Sola MC, Lim KC, Leach KM, Fandrey J, Ma Y, Knopfle G, Engel JD, Bungert J. Hepatic erythropoietin gene regulation by GATA-4. J Biol Chem. 2004 Jan 23;279(4):2955-61.

Epstein AC, Gleadle JM, McNeill LA, Hewitson KS, O'Rourke J, Mole DR, Mukherji M, Metzen E, Wilson MI, Dhanda A, Tian YM, Masson N, Hamilton DL, Jaakkola P, Barstead R, Hodgkin J, Maxwell PH, Pugh CW, Schofield CJ, Ratcliffe PJ. C. elegans EGL-9 and mammalian homologs define a family of dioxygenases that regulate HIF by prolyl hydroxylation. Cell. 2001 Oct 5;107(1):43-54.

Erslev AJ. Humoral regulation of red cell production. Blood 1953; 8: 349-357

Erslev AJ. In vitro production of erythropoietin by kidneys perfused with a serum-free solution. Blood. 1974 Jul;44(1):77-85

Fandrey J, Bunn HF. In vivo and in vitro regulation of erythropoietin mRNA: measurement by competitive polymerase chain reaction. Blood. 1993 Feb 1;81(3):617-23.

Fandrey J, Huwiler A, Frede S, Pfeilschifter J, Jelkmann W. Distinct signaling pathways mediate phorbol-ester-induced and cytokine-induced inhibition of erythropoietin gene expression. Eur J Biochem. 1994 Dec 1;226(2):335-40.

Fandrey J. Oxygen-dependent and tissue-specific regulation of erythropoietin gene expression. Am J Physiol Regul Integr Comp Physiol. 2004 Jun;286(6):R977-88.

Faquin WC, Schneider TJ, Goldberg MA. Effect of inflammatory cytokines on hypoxia-induced erythropoietin production. Blood. 1992 Apr 15;79(8):1987-94.

Fisher JW, Birdwell BJ. The production of an erythropoietic factor by the in situ perfused kidney. Acta Haematol. 1961;26:224-32.

Fried W. The liver as a source of extrarenal erythropoietin production. Blood. 1972 Nov;40(5):671-7

Ghezzi P, Brines M. Erythropoietin as an antiapoptotic, tissue-protective cytokine. Cell Death Differ. 2004 Jul;11 Suppl 1:S37-44.

Gordon AS, Cooper GW, Zanjani ED. The kidney and erythropoiesis. Semin Hematol. 1967 Oct;4(4):337-58.

Gurney CW, Goldwasser E, Pan C. Studies on erythropoiesis. VI. Erythropoietin in human plasma. J Lab Clin Med. 1957 Oct;50(4):534-42.

Heeschen C, Aicher A, Lehmann R, Fichtlscherer S, Vasa M, Urbich C, Mildner-Rihm C, Martin H, Zeiher AM, Dimmeler S. Erythropoietin is a potent physiologic stimulus for endothelial progenitor cell mobilization. Blood. 2003 Aug 15;102(4):1340-6.

Hellwig-Burgel T, Rutkowski K, Metzen E, Fandrey J, Jelkmann W. Interleukin-1beta and tumor necrosis factor-alpha stimulate DNA binding of hypoxia-inducible factor-1. Blood. 1999 Sep 1;94(5):1561-7.

Hill JM, Zalos G, Halcox JP, Schenke WH, Waclawiw MA, Quyyumi AA, Finkel T. Circulating endothelial progenitor cells, vascular function, and cardiovascular risk. N Engl J Med. 2003 Feb 13;348(7):593-600.

Imagawa S, Yamamoto M, Ueda M, Miura Y. Erythropoietin gene expression by hydrogen peroxide. Int J Hematol. 1996 Oct;64(3-4):189-95.

Ivan M, Kondo K, Yang H, Kim W, Valiando J, Ohh M, Salic A, Asara JM, Lane WS, Kaelin WG Jr. HIF alpha targeted for VHL-mediated destruction by proline hydroxylation: implications for O2 sensing. Science. 2001 Apr 20;292(5516):464-8.

Jacobs K, Shoemaker C, Rudersdorf R, Neill SD, Kaufman RJ, Mufson A, Seehra J, Jones SS, Hewick R, Fritsch EF, Isolation and characterization of genomic and cDNA clones of human erythropoietin. Nature. 1985 Feb 28-Mar 6;313(6005):806-10.

Jacobson LO, Goldwasser E, Fried W and Plazk L. Role of the kidney in erythropoiesis. Nature 1957; 179: 633-34

Jelkmann W and Hellwig-Burge. Biology of erythropoietin. Adv Exp Med Biol. 2001;502:169-87

Jelkmann W, Bauer C. Demonstration of high levels of erythropoietin in rat kidneys following hypoxic hypoxia. Pflugers Arch. 1981 Nov;392(1):34-9.

Jelkmann W. Erythropoietin. J Endocrinol Invest. 2003 Sep;26(9):832-7.

Jelkmann W. Erythropoietin: structure, control of production and function. Physiol Rev. 1992 Apr;72(2):449-89.

Juul SE, Yachnis AT, Christensen RD. Tissue distribution of erythropoietin and erythropoietin receptor in the developing human fetus. Early Hum Dev. 1998 Oct;52(3):235-49.

Kashii Y, Uchida M, Kirito K, Tanaka M, Nishijima K, Toshima M, Ando T, Koizumi K, Endoh T, Sawada K, Momoi M, Miura Y, Ozawa K, Komatsu N. A member of Forkhead family transcription factor, FKHRL1, is one of the downstream molecules of phosphatidylinositol 3-kinase-Akt activation pathway in erythropoietin signal transduction. Blood. 2000 Aug 1;96(3):941-9.

Kawakami M, Sekiguchi M, Sato K, Kozaki S, Takahashi M. Erythropoietin receptor-mediated inhibition of exocytotic glutamate release confers neuroprotection during chemical ischemia. J Biol Chem. 2001 Oct 19;276(42):39469-75.

Kieran MW, Perkins AC, Orkin SH, Zon LI. Thrombopoietin rescues in vitro erythroid formation from mouse embryos lacking the erythropoietin receptor. Proc Natl Acad Sci U S A 1996; 93: 9126–31

Kobayashi T, Yanase H, Iwanaga T, Sasaki R, Nagao M. Epididymis is a novel site of erythropoietin production in mouse reproductive organs. Biochem Biophys Res Commun. 2002 Aug 9;296(1):145-51.

Koury ST, Koury MJ, Bondurant MC, Caro J, Graber SE. Quantitation of erythropoietin-producing cells in kidneys of mice by in situ hybridization: correlation with hematocrit, renal erythropoietin mRNA, and serum erythropoietin concentration. Blood. 1989 Aug 1;74(2):645-51.

Kuratowska Z, Lewartowski B, Michalak E. Studies on the production of erythropoietin by isolated perfused organs. Blood. 1961 Nov;18:527-34.

Leverrier Y, Thomas J, Mathieu AL, Low W, Blanquier B, Marvel J. Role of PI3-kinase in Bcl-X induction and apoptosis inhibition mediated by IL-3 or IGF-1 in Baf-3 cells. Cell Death Differ. 1999 Mar;6(3):290-6.

Lin CH, Lim SK, D'Agati VG, Costantini F Differential effects of an erythropoietin receptor gene disruption on primitive and definitive erythropoiesis. Genes Dev. 1996 10:154–64

Lin FK, Suggs S, Lin CH, Browne JK, Smalling R, Egrie JC, Chen KK, Fox GM, Martin F, Stabinsky Z, et al. Cloning and expression of the human erythropoietin gene. Proc Natl Acad Sci U S A. 1985 Nov;82(22):7580-4.

Marti HH, Wenger RH, Rivas LA, Straumann U, Digicaylioglu M, Henn V, Yonekawa Y, Bauer C, Gassmann M. Erythropoietin gene expression in human, monkey and murine brain. Eur J Neurosci. 1996 Apr;8(4):666-76.

Masuda S, Kobayashi T, Chikuma M, Nagao M, Sasaki R. The oviduct produces erythropoietin in an estrogen- and oxygen-dependent manner. Am J Physiol Endocrinol Metab. 2000 Jun;278(6):E1038-44.

Maxwell PH, Wiesener MS, Chang GW, Clifford SC, Vaux EC, Cockman ME, Wykoff CC, Pugh CW, Maher ER, Ratcliffe PJ. The tumour suppressor protein VHL targets hypoxia-inducible factors for oxygen-dependent proteolysis. Nature. 1999 May 20;399(6733):271-5.

Miescher F. Über die beziehungen zwischen meereshöhe und beschaffenheit des blutes. Corresp. Bl. Schweiz Aerzte 1893; 23: 809-830

Miller CP, Liu ZY, Noguchi CT, Wojchowski DM. A minimal cytoplasmic subdomain of the erythropoietin receptor mediates erythroid and megakaryocytic cell development. Blood. 1999 Nov 15;94(10):3381-7.

Miyake T, Kung CK, Goldwasser E. Purification of human erythropoietin. J Biol Chem. 1977 Aug 10;252(15):5558-64

Morishita E, Masuda S, Nagao M, Yasuda Y, Sasaki R. Erythropoietin receptor is expressed in rat hippocampal and cerebral cortical neurons, and erythropoietin prevents in vitro glutamate-induced neuronal death. Neuroscience. 1997 Jan;76(1):105-16.

Reissmann KR. Studies on the mechanism of erythropoietic stimulation in parabiotic rats during hypoxia. Blood 1950; 5: 372-380

Ribatti D, Presta M, Vacca A, Ria R, Giuliani R, Dell'Era P, Nico B, Roncali L, Dammacco F. Human erythropoietin induces a pro-angiogenic phenotype in cultured endothelial cells and stimulates neovascularization in vivo. Blood 1999;93:2627 –2636

Salceda S, Caro J. Hypoxia-inducible factor 1alpha (HIF-1alpha) protein is rapidly degraded by the ubiquitin-proteasome system under normoxic conditions. Its stabilization by hypoxia depends on redox-induced changes. J Biol Chem. 1997 Sep 5;272(36):22642-7.

Saray A, Ozakpinar R, Koc C, Serel S, Sen Z, Can Z. Effect of chronic and short-term erythropoietin treatment on random flap survival in rats: an experimental study. Laryngoscope. 2003 Jan;113(1):85-9.

Schuster SJ, Wilson JH, Erslev AJ, Caro J. Physiologic regulation and tissue localization of renal erythropoietin messenger RNA. Blood. 1987 Jul;70(1):316-8.

Shi Q, Rafii S, Wu MH, Wijelath ES, Yu C, Ishida A, Fujita Y, Kothari S, Mohle R, Sauvage LR, Moore MA, Storb RF, Hammond WP. Evidence for circulating bone marrow-derived endothelial cells. Blood. 1998 Jul 15;92(2):362-7.

Sieff CA, Ekern SC, Nathan DG, Anderson JW. Combinations of recombinant colony-stimulating factors are required for optimal hematopoietic differentiation in serum-deprived culture. Blood. 1989 Feb 15;73(3):688-93.

Sieff CA, Emerson SG, Mufson A, Gesner TG, Nathan DG. Dependence of highly enriched human bone marrow progenitors on hemopoietic growth factors and their response to recombinant erythropoietin. J Clin Invest. 1986 Jan;77(1):74-81.

Silva M, Benito A, Sanz C, Prosper F, Ekhterae D, Nunez G, Fernandez-Luna JL. Erythropoietin can induce the expression of bcl-x(L) through Stat5 in erythropoietin-dependent progenitor cell lines. J Biol Chem. 1999 Aug 6;274(32):22165-9.

Siren AL, Knerlich F, Poser W, Gleiter CH, Bruck W, Ehrenreich H. Erythropoietin and erythropoietin receptor in human ischemic/hypoxic brain. Acta Neuropathol (Berl). 2001 Mar;101(3):271-6.

Socolovsky M, Fallon AE, Wang S, Brugnara C, Lodish HF. Fetal anemia and apoptosis of red cell progenitors in Stat5a-/-5b-/- mice: a direct role for Stat5 in Bcl-X(L) induction. Cell. 1999 Jul 23;98(2):181-91.

Spivak JL, Hogans BB. The in vivo metabolism of recombinant human erythropoietin in the rat. Blood 1989; 73:90–9

Spivak JL, Pham T, Isaacs M, Hankins WD. Erythropoietin is both a mitogen and a survival factor. Blood. 1991 Mar 15;77(6):1228-33.

Suda T, Suda J, Ogawa M. Disparate differentiation in mouse hemopoietic colonies derived from paired progenitors. Proc Natl Acad Sci U S A. 1984 Apr;81(8):2520-4.

Tramontano AF, Muniyappa R, Black AD, Blendea MC, Cohen I, Deng L, Sowers JR, Cutaia MV, El-Sherif N. Erythropoietin protects cardiac myocytes from hypoxia-induced apoptosis through an Akt-dependent pathway. Biochem Biophys Res Commun. 2003 Sep 5;308(4):990-4.

Vasa M, Fichtlscherer S, Aicher A, Adler K, Urbich C, Martin H, Zeiher AM, Dimmeler S. Number and migratory activity of circulating endothelial progenitor cells inversely correlate with risk factors for coronary artery disease. Circ Res. 2001 Jul 6;89(1):E1-7.

Viault F. Sur l'augmentation considérable du nombre des globules rouge dans le sang chez les inhabitants des hauts plateaux de l'Amerique du Sud. C R Acad Sci Paris 1890; 111:917-918

Wang GL, Jiang BH, Rue EA, Semenza GL. Hypoxia-inducible factor 1 is a basic-helix-loop-helix-PAS heterodimer regulated by cellular O2 tension. Proc Natl Acad Sci U S A. 1995 Jun 6;92(12):5510-4.

Wojchowski DM, Gregory RC, Miller CP, Pandit AK, Pircher TJ. Signal transduction in the erythropoietin receptor system. Exp Cell Res. 1999 Nov 25;253(1):143-56.

Wu H, Lee SH, Gao J, Liu X, Iruela-Arispe ML. Inactivation of erythropoietin leads to defects in cardiac morphogenesis. Development. 1999 Aug;126(16):3597-605.

Yokomizo R, Matsuzaki S, Uehara S, Murakami T, Yaegashi N, Okamura K. Erythropoietin and erythropoietin receptor expression in human endometrium throughout the menstrual cycle. Mol Hum Reprod. 2002 May;8(5):441-6.

Zanjani ED, Peterson EN, Gordon AS, Wasserman LR. Erythropoietin production in the fetus: role of the kidney and maternal anemia. J Lab Clin Med. 1974 Feb;83(2):281-7

Chapter 2

EXPRESSION OF ERYTHROPOIETIN AND ITS RECEPTOR IN THE CENTRAL NERVOUS SYSTEM

[1]Hugo H. Marti and [2]Christian Bauer

[1]Institut für Physiologie und Pathophysiologie, Universität Heidelberg, Im Neuenheimer Feld 326, D-69120 Heidelberg, Germany; [2]Physiologisches Institut, Universität Zürich, Winterthurerstrasse 190, CH-8057 Zürich, Switzerland

Abstract: Erythropoietin (EPO) is a glycoprotein that is produced mainly by interstitial fibroblasts in the kidney. Released into the circulation, EPO makes its way to the bone marrow where it regulates red cell production by preventing apoptosis of erythroid progenitor cells. Recently, EPO has emerged as a multifunctional growth factor that plays a significant role in the nervous system. Both EPO and its receptor are expressed throughout the brain in glial cells, neurons and endothelial cells. Brain-derived EPO is upregulated by hypoxia, and expression of both EPO and its receptor are specifically modulated during cerebral ischemia. EPO has potent neuroprotective properties *in vivo* and *in vitro* and appears to act in a dual way by directly protecting neurons from ischemic damage and by stimulating endothelial cells, and thus supporting the growth of new blood vessels. EPO eventually also modulates inflammatory responses. Thus, hypoxically upregulated EPO is a naturally self-regulated physiological protective mechanism in the mammalian brain, especially during ischemia. As EPO is also a clinically extremely well studied and tolerated compound, its use in stroke patients is tempting.

Key words: neuron; astrocyte; microglia; endothelial cell; HIF-1; VEGF; IGF-1; angiogenesis; neuroprotection; apoptosis; hypoxia; ischemia; stroke.

1. INTRODUCTION

Clinicians, who are treating patients suffering from the anemia of End-Stage Renal Failure with recombinant erythropoietin (EPO), have often reported an improvement of the cognitive function of their patients (reviewed by Ehrenreich and Siren, 2001a). However, it was never quite clear if this enhancement of cognitive function is the result of an increase in the oxygen transport capacity of the blood leading to an improved oxygenation of the brain, or if EPO has a direct effect on brain cells by itself. This latter assumption was largely disregarded because EPO was not thought to cross the blood-brain barrier, due to its large size (30 kDa) and its many negative charges (Davis et al., 1987;Recny et al., 1987). Indeed, results from various studies indicated that endogenous kidney-derived EPO only gets access into the brain after breakdown of the blood-brain barrier (Marti et al., 1997;Buemi et al., 2000). Nonetheless, recent evidence suggests that high amounts of recombinant EPO can attain the brain in a number of experimental settings (Brines et al., 2000;Juul et al., 2004), a finding that may explain the beneficial effects on cognitive function seen in these patients.

These observations led to two interesting questions that concern the presence of EPO and EPO receptors in the brain. First, as EPO mediates its effects through binding to its cognate receptor, EPO receptor should be expressed at the site of action in the central nervous system (CNS) to enable EPO to elicit biological functions. Second, if EPO receptors are naturally occurring in the CNS, one has to postulate that EPO is endogenously produced in the brain itself to activate these receptors, on the assumption that kidney-derived EPO does not cross the blood-brain barrier under physiological conditions. Indeed, it was demonstrated that EPO receptors are widely distributed in the mammalian brain, and that the expression of EPO mRNA and EPO protein largely coincides with the occurrence of EPO receptor mRNA and protein (reviewed by Marti and Bernaudin, 2003;Genc et al., 2004;Marti, 2004). The upregulation of brain EPO in a large number of experimental conditions associated with tissue hypoxia is well documented, and comprise many mammalian species including mice, rats, monkeys, and humans (Marti et al., 1996;Marti et al., 2000;Siren et al., 2001;Genc et al., 2004).

In this chapter we will first deal with the normal pO_2 gradients within the brain. We will next consider those areas of the CNS that produce EPO and carry EPO receptors, and will finally have a look at the physiology of EPO function in the CNS, as an actor on neurons, glial cells and endothelial cells, including the action of EPO on the architecture of brain vessels.

2. OXYGEN GRADIENTS WITHIN THE BRAIN AND EXPRESSION OF HIF-1α AND HIF-2 α

The brain exhibits a high rate of oxygen consumption, comprising some 20 % of the normal oxygen consumption at rest. Within the brain, the pO2 profiles are low and non-uniform, ranging from 2 mmHg in the fornical part of the pons all the way up to 25-35 mm Hg in the cerebral cortex and 20 mmHg in the hippocampus (reviewed by Erecinska and Silver, 2001). The brain is therefore particularly sensitive towards hypoxia. Accordingly, a marked hypoxic induction of the transcription factors HIF-1 (hypoxia-inducible factor 1) (Bergeron et al., 1999) and HIF-2 (Wiesener et al., 2003) in the brain can be readily observed. Both HIF-1 and HIF-2 are dimers, having in common a constitutively expressed β-chain, yet their individual α-subunits (HIF-1 α, HIF-2 α) are inducible, e.g. by hypoxia.

These transcription factors direct many hypoxic responses such as the production of EPO, the Vascular Endothelial Growth Factor (VEGF), and the glucose transporter Glut-1. Classically, the regulation of these transcription factors occurs through an immediate oxygen-dependent proteolysis of their respective alpha subunits, HIF-1α and HIF-2α (reviewed by Schofield and Ratcliffe, 2004;Sharp and Bernaudin, 2004). HIF-1α is ubiquitously expressed, in the brain especially in parenchymal cells like the neurons, whilst HIF-2α is more confined to non-parenchymal cells i.e., endothelial and glial cells (Wiesener et al., 2003). These findings indicate that redundancy in the induction and thus function of HIF-1α and of HIF-2α is limited and that both factors may have complementary functions in a coordinated transcriptional response to hypoxia.

The notion for an oxygen-dependent regulation of HIF-1α and HIF-2α holds certainly true for a more immediate hypoxic reactions, i.e. minutes to two days, as compared to conditions of global brain hypoxia lasting more than one week. Under such chronic conditions, the Insulin-Like Growth Factor 1 (IGF-1) apparently adopts the role of a signal that keeps HIF-1α (and possibly HIF-2α) elevated despite the fact that the hypoxic marker EF5 does not indicate tissue hypoxia (Chavez and LaManna, 2002). This upregulation of HIF-1α by IGF-1 is accompanied by an induction of some of the classical HIF-1 target genes e.g. EPO, VEGF and Glut-1 in the brain (Chavez and LaManna, 2002). In keeping with these observations are the results obtained by Masuda and co-workers (Masuda et al., 1997) who assessed a hypoxia-independent stimulation of EPO mRNA by IGF-1 in primary cultures of brain astrocytes.

All in all, these results strongly suggest that IGF-1 and hypoxia enhance two distinct and independent mechanisms of activation of HIF-1 and its targets genes, and may thereby account for the neuronal rescue by IGF-1 found in conditions of post-ischemic brain injury (Guan et al., 2003). This assumption holds true, despite of the fact that the induction of HIF-1α by IGF-1 has not yet been clearly elucidated at the molecular level. In addition, recent results, demonstrating synergistic cooperative actions of EPO and IGF-1 that confer acute neuroprotection on cultured neurons indicate yet another level of complex interaction between theses two cytokines (Digicaylioglu et al., 2004).

3. EXPRESSION OF EPO AND EPO RECEPTORS IN THE CNS

It came to a surprise, when it was found that both EPO and EPO receptor are constitutively expressed in the brain of rats and mice (Tan et al., 1992;Digicaylioglu et al., 1995;Marti et al., 1996), as it was largely held that the two entities of the "EPO System" are disconnected, one residing in the kidney (the EPO production machinery), and the other one in the bone marrow (the EPO receptor machinery). It soon turned out that these two entities of the EPO system are located within one and the same organ, even within one single cell, not only in the brain of rodents, but in the brain of monkeys and humans as well (Marti et al. 1996) (Fig. 2-1).

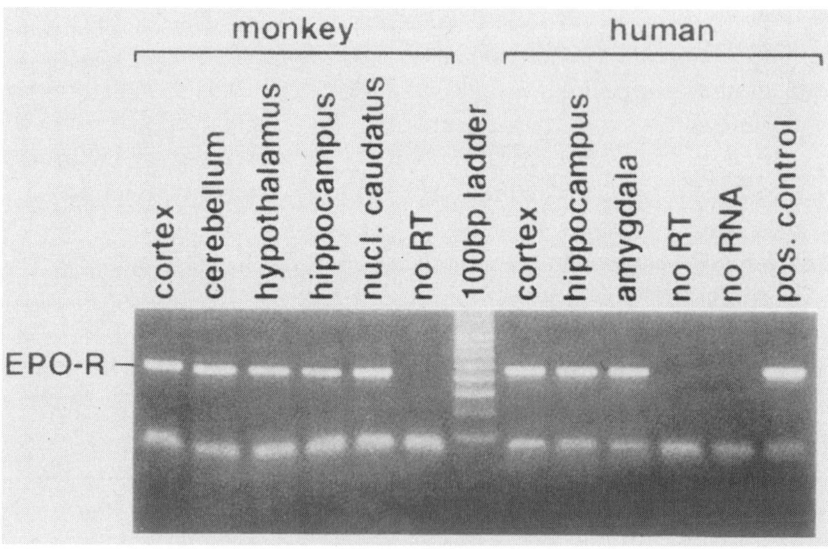

Figure 2-1. EPO receptor mRNA in different regions of the monkey and human brain. One microgram of total RNA was analyzed by RT-PCR specific for primate EPO receptor (EPOR). The RT-PCR product for primate EPO receptor is 485 bp. No RT, no reverse transcriptase added; positive control, human EPO receptor cDNA. (Reproduced by permission of Blackwell Publishing Ltd. from Marti et al., 1996).

In the monkey brain EPO expression was detected in all areas investigated, including various cortical areas, the cerebellum, the hypothalamus, the hippocampus, and the caudate nucleus. In the few samples of the human brain that became available at the time, EPO mRNA along with immunoreactive EPO was found in the temporal cortex and the hippocampus (Marti et al., 1996;Siren et al., 2001). The detection of the "EPO System", *i.e.* EPO and its receptor, in the human brain immediately attracted a broader attention, particularly in view of the fact that EPO also occurs in the cerebrospinal fluid of human adults (Marti et al., 1997) and neonates (Juul et al., 1997).

The next question concerns the cell type(s) that produce EPO, and carry EPO receptors. It was soon detected that not only astrocytes that are in close contact both with neurons and brain capillaries produce EPO (Masuda et al., 1994;Marti et al., 1996) but that neuronal cells themselves are a source of brain-derived EPO, as was shown both *in vitro* and *in vivo* (Bernaudin et al., 1999;Bernaudin et al., 2000;Siren et al., 2001). Expression of EPO receptor mRNA and protein was demonstrated in the brain of mouse, rat, monkey and humans (Digicaylioglu et al., 1995;Marti et al., 1996;Liu et al., 1997), and the apposite analysis at the cellular level revealed that neurons and astrocytes carry the EPO receptor (Bernaudin et al., 1999;Bernaudin et al., 2000;Siren et al., 2001). In addition, it was shown that neurons, astrocytes and the microglia of human origin express EPO receptor mRNA in primary cultures of the respective cell type (Nagai et al., 2001). Of particular note is the fact that yet another glia cell type, the oligodendrocytes, was found to be devoid of EPO mRNA and of EPO receptor mRNA in purified cultures of adult human origin whilst immature oligodendrocytes isolated from embryonic rat brain were shown to produce EPO, as well as its receptor (Nagai et al., 2001).

Apart from neurons and glial cells, there is yet another cell type that deserves mention. These are the vascular endothelial cells that are one of the targets for EPO in the CNS, and suggest therefore a role for EPO in the vascular response to brain injury. This notion is supported by the fact that a strong immunoreactivity of the EPO receptor was found to be associated with brain endothelial cells (Brines et al., 2000), and that the astrocytic processes that surround brain capillaries carry a dense population of the EPO receptor (Brines et al., 2000). These *in vivo* findings are supported by *in vitro* analyses of endothelial cells derived from the human umbilical cord (Anagnostou et al., 1994) and rat brain microvessels (Yamaji et al., 1996) which clearly demonstrated that EPO receptors are also associated with brain endothelial cells. All of these findings are compatible with the idea that neuronal cells, glial cells and endothelial cells form an interactive meshwork that promotes the formation of new blood vessels in the brain in situations of an imminent ischemia.

In summary, EPO as well as EPO receptors are expressed in various parts of the CNS. **Table 2-1** presents an overview on the cellular sites of EPO and EPO receptor expression in the CNS.

Table 2-1. Cellular expression of EPO and its receptor in the CNS of rodents and men

Protein	EPO				EPOR			
	Rodents		Humans		Rodents		Humans	
	In vitro	*In vivo*	*In vitro*	*In vivo*	*In vitro*	*In vivo*	*In vitro*	*In vivo*
Neurons	+	+	-	+	+	+	+	+
Astro	+	+	+	+	+	+	+	+
Micrgl	+	n.a.	-	n.a.	n.a.	n.a.	+	n.a.
Oligodc	n.a.	n.a.	-	n.a.	+	n.a.	-	n.a.
Ec	-	?	n.a.	?	+	+	+	+

Astro, astrocytes; Micrgl; microglia; Oligodc, oligrodendrocytes; Ec, endothelial cells; +, expression detected; -, no expression; n.a., not analysed; ?, not proven. (Adapted from Marti, 2004).

We will now regard in more detail the "EPO Network" in the brain taking in consideration neurons, glial cells and brain vascular endothelial cells.

4. NEURONAL CELLS: PHYSIOLOGY OF EPO

A high expression of both EPO and EPO receptors on neurons is found in those areas of the brain that are known to be particularly susceptible towards acute hypoxia (Lipton, 1999), e.g. the telencephalon (endbrain) and the hippocampus (Digicaylioglu et al., 1995). This exceptional sensitivity towards hypoxic insult is reflected by the high expression of synapses that use glutamate as a transmitter in those areas of the brain that are concerned with the storage and the retrieval of memory contents (Tang et al., 1999;Miu et al., 2004). In view of the fact that EPO protects neuronal cells from glutamate "overflow " (Morishita et al., 1997), the following scheme may be envisioned regarding the physiological counterbalance between glutamate and EPO in the brain: **i**) EPO decreases the number and/or the sensitivity of glutamate receptors of the NMDA type on the extracellular surface of neuronal cells, **ii**) EPO decreases glutamate release from neurons, **iii**) EPO increases glutamate uptake in presynaptic terminals and astrocytes, and **iv**) EPO increases the release of γ-amino butyric acid (GABA) that has an inhibitory effect on the postsynaptic structures of those neurons that use glutamate as a transmitter. Future research will show which of these possibilities is effective. In this way, EPO may be regarded as a "natural" modifier of glutamate action that helps to physiologically buffer the excitatory action of glutamate, thereby preventing the occurrence of a local glutamate excitotoxicity, and/or EPO may also contribute to a modification of the NMDA receptor-mediated consolidation of memory contents (Tang et al., 1999).

All of the neuronal structures mentioned above carry EPO receptors that are necessary and sufficient to transmit the EPO "signal" into the cell. It is interesting to note in this connection that EPO is involved in more complex neurological functions like synaptic transmission (Weber et al., 2002), better performance in pain and fear paradigms (Campana and Myers, 2003;Miu et al., 2004), in cognitive function (Ehrenreich and Siren, 2001a) and in spatial navigation performance (Sadamoto et al., 1998). There is evidence that EPO stimulates neuronal function and viability via activation of calcium channels (Assandri et al., 1999) and release of neurotransmitters (Koshimura et al., 1999;Yamamoto et al., 2002). Although the precise molecular mechanisms that account for these findings are so far poorly understood, it has become clear that EPO possesses a clear-cut neurotrophic activity: EPO augments choline acetyltransferase activity in primary cultured mouse septal neurons and promotes regeneration of septal cholinergic neurons in adult rats which had undergone fimbria-fornix transections (Konishi et al., 1993). Recently, it was also demonstrated that addition of EPO enhances survival and dopaminergic differentiation of CNS precursor cells *in vitro* (Studer et al., 2000) reminiscent of its survival function during erythropoiesis in the bone marrow. Furthermore, hypoxia-induced EPO appears to act directly on forebrain neural stem cells, promoting the production of neuronal progenitors suggesting that EPO is involved in neurogenesis after hypoxia (Shingo et al., 2001). All these results suggest that EPO is a neurotrophic factor that supports both differentiation and growth of neurons during development.

It should also be noted that not only the neuronal network in the brain profits from locally released EPO, but that the same notion is valid for the spinal cord. In this part of the CNS it was shown that α-motoneurons of the spinal cord benefit from the application of EPO in conditions of a spinal cord ischemic injury by the very same mechanisms that were described for the more centrally located neurons (Celik et al., 2002).

So far, we have mainly considered the interaction of the "EPO system" with neuronal cells, and will now turn to a population of cells that outnumbers the neurons by a factor 10, namely the glial cells.

5. GLIAL CELLS: DEFINITIONS AND CONCEPTS REGARDING EPO

The term "glia" is derived from the Greek word *glia*, which literally means "glue", and indeed the glial cells have long been regarded as a cellular glue that hold the neuronal elements together. No statement could be less true in regard to a cell population that has very diverse physiological functions, as glia cells are involved in virtually every aspect of neural function (Zhang, 2001).

Let us now consider these functions in more detail: Largely the glial cells can be divided in microglia and macroglia that have diverse functions. The microglia is concerned with the recognition and processing of foreign antigens, and would thereby qualify as an immune competent cell type that identifies any foreign material entering the brain. In addition, the microglia has interesting functions in the recruitment of neuronal stem cells (Zhang, 2001;Doetsch, 2003). Considering that EPO receptor expression was shown in cultures of human microglial cells (Nagai et al., 2001), it remains a challenging task for future research activities if EPO can "convince" the microglia to recruit ever more neuronal stem cells. Furthermore, EPO can act as an anti-inflammatory cytokine during cerebral ischemia by reducing the recruitment of microglia into the infarcted area (Villa et al., 2003) and, thus, play an immunomodulatory role in the CNS.

The macroglia on the other hand, may be divided in two major cell types: the star-shaped astrocytes, the term being derived from the Greek word *astron*, meaning star, and the oligodendrocytes from *oligos*, meaning few. Oligodendrocytes are myelin-forming cells that wrap the axons, and provide the basis for a rapid transmission of action potentials from the central to the peripheral part of the CNS. In certain diseases, this function is lost, for example in multiple sclerosis. Under the assumption that EPO promotes growth and differentiation of embryonic oligodendrocytes, the hypothesis can be put forward that the "EPO system" supports the myelinization of the growing axons in the embryonic CNS (Sugawa et al., 2002). This idea tallies nicely with the developmental regulation of EPO and its receptor in the embryonic and fetal brain of animals and humans, and suggests therefore a widespread role of the "EPO system" in the maturation of the mammalian CNS (Juul, 2002).

Astrocytes on the other hand provide important structural, metabolic and trophic support to neurons. They encase the brain, interact with endothelial cells to form the blood-brain barrier, absorb neurotransmitters, maintain extracellular ion homeostasis and secrete growth factors, cytokines and components of the extracellular matrix. Astrocytes are further involved in the formation and stabilization of synapses and the modulation of synaptic efficacy (Doetsch, 2003). Despite of their functional importance, the experimental evidence obtained so far - regarding the effects of the "EPO system" on the more specific function of glial cells - is pretty sparse. Nonetheless, EPO was shown to effectively support the development of astrocytes in a concentration-dependent manner (Sugawa et al., 2002). In this way, astrocytes may thus be physiologically involved in the EPO-induced modulation of excitatory postsynaptic glutamate responses as well (Kawakami et al., 2001).

None of these functions in neurons or glial cells can be maintained without an appropriate blood flow feeding these cells with oxygen and nutrients. Ideally, regulation of blood flow should thus also be geared by EPO. Recent insights suggest that blood vessels and nerves have much more in common than was originally anticipated (reviewed by Carmeliet, 2003). They use similar principles to differentiate, grow and navigate towards their targets. Moreover, the vascular and the nervous systems cross-talk by using similar signals and factors. These factors include VEGF, angiopoietins, ephrins, platelet-derived growth factors, and EPO. Therefore, we will next consider the maintenance of blood vessels in the brain that are under the direction of EPO and other oxygen-controlled cytokines.

6. EPO AND THE ARCHITECTURE OF BRAIN VESSELS

Formation and remodeling of new blood vessels (vascular morphogenesis) is governed by three processes, vasculogenesis, angiogenesis and arteriogenesis. **Vasculogenesis** is defined as the differentiation of mesodermal progenitor cells (angioblasts) into endothelial cells in situ where they subsequently aggregate and form a primary vascular plexus (Risau and Flamme, 1995). Vasculogenesis occurs primarily during embryonic development. **Angiogenesis** is defined as the formation of new blood vessels by sprouting of endothelial cells from pre-existing vessels or by intravascular subdivision (intussusception) (Risau, 1997). Angiogenesis further refines the primitive embryonic vascular plexus and includes remodeling, a process that transforms the relatively uniformly sized vasculature into the network of small and large vessels, that finally undergoes maturation by recruiting perivascular cells, such as smooth muscle cells and pericytes. Angiogenesis is an important process during embryogenesis but also occurs in the adult in response to altered metabolic requirements, e.g. it can be triggered by hypoxia. Finally, **arteriogenesis** is the rapid proliferation of pre-existing collateral vessels that occurs in ischemic tissue (for a review see Buschmann and Schaper, 1999). EPO could be involved in all three processes, acting on endothelial cells, vascular smooth muscle cells and even cardiomyocytes (reviewed by Smith et al., 2003).

Originally, EPO was shown to act on the mitogenesis and chemotaxis of endothelial cells derived from the human umbilical vein and bovine adrenal capillaries (Anagnostou et al., 1994). It was then shown that vessel outgrowth of rat aortic rings was stimulated by EPO (Carlini et al., 1995) suggesting that EPO has angiogenic properties. Indeed, neovascularization *in vivo* is stimulated in the endometrium after EPO injection into the mouse uterine cavity (Yasuda et al., 1998). In the brain, EPO induces a dose-dependent mitogenic activity on brain capillary endothelial cells (Yamaji et al., 1996).

With regard to endothelial functions of EPO, it is noteworthy that endothelial cells and hematopoietic cells are derived from the same mesenchymal precursor, the so-called hemangioblast (Risau, 1997). This may explain why endothelial cells carry the EPO receptor and can be stimulated by EPO (Ribatti et al., 1999;Yasuda et al., 1998). Very recently, it was shown that EPO is a potent physiologic stimulus for endothelial progenitor cell mobilization and EPO stimulates postnatal neovascularization (Heeschen et al., 2003). In erythroid precursor cells and neurons, EPO is a survival factor. The same notion seems to hold true for endothelial cells; EPO prevents cell injury and DNA fragmentation (Chong et al., 2002). EPO might also influence endothelial cells indirectly through the activation of the VEGF/VEGF receptor system. Interestingly, it was shown that EPO-induced proliferation of bovine aortic and glomerular endothelial cells was prevented by a specific anti-VEGF antibody (Nitta et al., 1999;Victoria et al., 1998). Thus, EPO might exert its function on endothelial cells via activation of VEGF receptors. Indeed, mRNA expression for both the VEGF receptor-1 and VEGF receptor-2 was upregulated in aortic cells after EPO pretreatment (Victoria et al., 1998).

It thus appears likely that increased expression of EPO and its receptor in blood vessels during cerebral ischemia in mice (Bernaudin et al., 1999) as well as in humans (Siren et al., 2001) contributes to new vessel growth in the tissue area suffering from hypoxia. Hypoxia- or ischemia-induced EPO thus stimulates new vessel growth enabling the transport of more red blood cells and thereby increasing the amount of oxygen delivered to the hypoxic tissue which in turn counteracts the detrimental effects of hypoxia on neurons.

7. PATHOPHYSIOLOGY OF EPO: CONTROLLING THE AFTERMATH OF TISSUE HYPOXIA

EPO has been shown to be neuroprotective in the brain after exposure to a variety of insults, including cerebral ischemia, head injury, seizures and experimental autoimmune encephalomyelitis (reviewed in Marti, 2004). In 1998, Sasaki and colleagues provided good evidence that endogenous brain-derived EPO is crucial for neuronal survival *in vivo*. Infusion of a soluble EPO receptor into the brain of gerbils, submitted to a mild ischemia that did not produce neuronal damage by itself, resulted in neuronal cell death in the hippocampus (Sakanaka et al., 1998). These results clearly indicate that brain-derived EPO is an endogenous protective agent for neurons against mild forms of tissue hypoxia and ischemia. Thus, the "EPO system" serves as an endogenous system to protect brain cells from damage caused by intermittent episodes of hypoxia. Along this line, EPO has been implicated in the mechanisms of ischemic tolerance or preconditioning. Preconditioning means that practically any stimulus capable of causing injury to a tissue can, when applied below the threshold level of damage, activate endogenous protective mechanisms and thus potentially lessen the impact of subsequent, more severe insults (reviewed in Dirnagl et al., 2003). In models of ischemic preconditioning both *in vitro* and *in vivo*, hypoxia-induced EPO release from astrocytes can inhibit hypoxia-induced apoptosis in neurons (Ruscher et al., 2002) and thus provide stroke tolerance (Prass et al., 2003).

The fact that a differential temporal and cellular modulation of the "EPO system" by ischemia is also detected in human brain tissue (Siren et al., 2001) indicates that EPO may have a beneficial effect for the treatment of stroke patients. Results from a first clinical phase I / phase II study are indeed promising. Intravenous high-dose recombinant EPO in a total of 53 stroke patients was well tolerated and associated with an improvement in clinical outcome at one month (Ehrenreich et al., 2002). Currently, a larger multi centre study is under way.

8. SUMMARY

A primary goal, not only for stroke patients but also for other diseases of the CNS, is to protect neural function. Imitation of brain endogenous protective mechanisms may be the key to future successful approaches to neuroprotection; therefore activation and mimicry of endogenous mechanisms can be expected to be efficient and well tolerated (Ehrenreich and Siren, 2001b). In this respect, EPO may be a showpiece. Originally identified as a hematopoietic factor, EPO is expressed in the CNS, including the human brain. Brain-derived EPO is upregulated by hypoxia and IGF-1, and expression of both EPO and EPO receptor are specifically modulated during cerebral ischemia. EPO has a neuroprotective potential both *in vitro* and *in vivo* in various animal models of CNS diseases by inhibition of apoptosis in neurons and inducing angiogenesis. EPO eventually also modulates inflammatory responses (Fig. 2-2). Thus, hypoxically upregulated EPO is a naturally self-regulated physiological protective mechanism in the mammalian brain, especially during ischemia. As EPO is also a clinically extremely well studied and tolerated compound, its use in stroke patients is tempting.

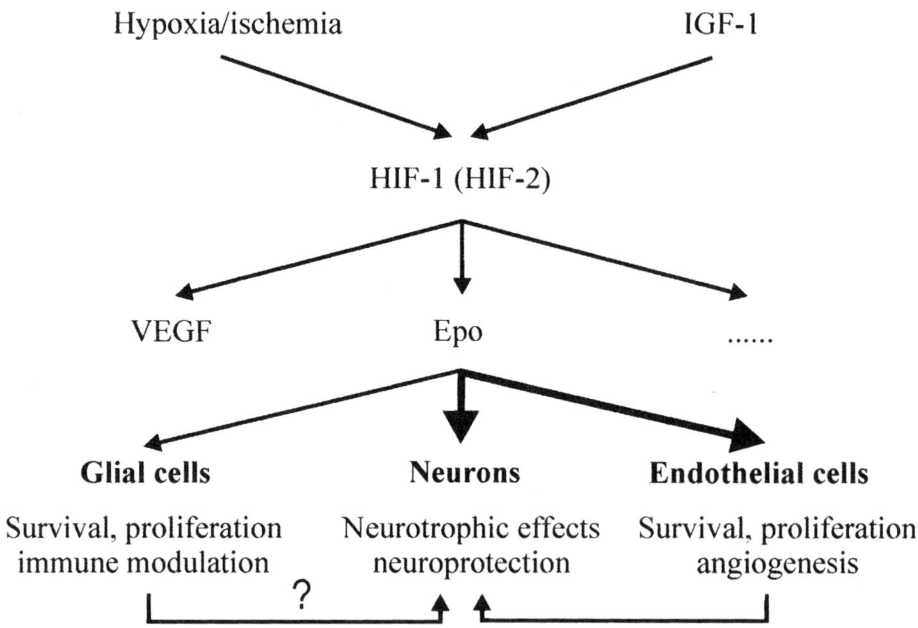

Figure 2-2. Schematic diagram of the possible actions of EPO in the CNS. Tissue hypoxia, cerebral ischemia, as well as the Insulin-like Growth Factor-1 (IGF-1) activate the hypoxia-inducible factor-1 (HIF-1) (and possibly HIF-2), that in turn activates gene transcription of a variety of oxygen-regulated factors, among them erythropoietin (EPO) and vascular endothelial growth factor (VEGF). The main targets (characterized by thick arrows) for EPO are neurons and endothelial cells, conferring cellular protection. The prevention of apoptosis and stimulation of proliferation of endothelial cells, results in new vessel growth (angiogenesis), and ultimately better oxygenation of hypoxic tissues. In addition, the EPO receptor is expressed on microglial cells and astrocytes, making glial cells to an additional target for EPO, although the effects on these cells are less clear and the contribution to neuronal survival remains to be established. (Adapted from Marti, 2004).

All these results support the idea that EPO acts in the CNS by a variety of mechanisms in neurons, endothelial cells and glial cells. Stimulation of new vessel growth (angiogenesis) leads to a better tissue oxygenation in the brain, in addition to its erythropoietic effect resulting in an increased oxygen carrying capacity of the blood. EPO also modulates electrophysiological and inflammatory responses, making it to an attractive neurotrophic and neuroprotective factor. Thus – coming back to the initial clinical observation - cognitive function might be indeed improved by a direct action of EPO within the CNS.

REFERENCES

Anagnostou A, Liu Z-Y, Steiner M, Chin K, Lee ES, Kessimian N, Noguchi CT (1994) Erythropoietin-receptor mRNA expression in human endothelial cells. Proc Natl Acad Sci USA 91:3974-3978.

Assandri R, Egger M, Gassmann M, Niggli E, Bauer C, Forster I, Görlach A (1999) Erythropoietin modulates intracellular calcium in a human neuroblastoma cell line. J Physiol (London) 516:343-352.

Bergeron M, Yu AY, Solway KE, Semenza GL, Sharp FR (1999) Induction of hypoxia-inducible factor-1 (HIF-1) and its target genes following focal ischaemia in rat brain. Eur J Neurosci 11:4159-4170.

Bernaudin M, Marti HH, Roussel S, Divoux D, Nouvelot A, MacKenzie ET, Petit E (1999) A potential role for erythropoietin in focal permanent cerebral ischemia in mice. J Cereb Blood Flow Metab 19:643-651.

Bernaudin M, Bellail A, Marti HH, Yvon A, Vivien D, Duchatelle I, MacKenzie ET, Petit E (2000) Neurons and astrocytes express EPO mRNA: oxygen-sensing mechanisms that involve the redox-state of the brain. Glia 30:271-278.

Brines ML, Ghezzi P, Keenan S, Agnello D, de Lanerolle NC, Cerami C, Itri LM, Cerami A (2000) Erythropoietin crosses the blood-brain barrier to protect against experimental brain injury. Proc Natl Acad Sci USA 97:10526-10531.

Buemi M, Allegra A, Corica F, Floccari F, D'Avella D, Aloisi C, Calapai G, Iacopino G, Frisina N (2000) Intravenous recombinant erythropoietin does not lead to an increase in cerebrospinal fluid erythropoietin concentration. Nephrol Dial Transplant 15:422-423.

Buschmann I, Schaper W (1999) Arteriogenesis versus angiogenesis: two mechanisms of vessel growth. News Physiol Sci 14:121-125.

Campana WM, Myers RR (2003) Exogenous erythropoietin protects against dorsal root ganglion apoptosis and pain following peripheral nerve injury. Eur J Neurosci 18:1497-1506.

Carlini RG, Reyes AA, Rothstein M (1995) Recombinant human erythropoietin stimulates angiogenesis in vitro. Kidney Int 47:740-745.

Carmeliet P (2003) Blood vessels and nerves: common signals, pathways and diseases. Nature Rev Genet 4:710-720.

Celik M, Gökmen N, Erbayraktar S, Akhisaroglu M, Konakc S, Ulukus C, Genc S, Genc K, Sagiroglu E, Cerami A, Brines M (2002) Erythropoietin prevents motor neuron apoptosis and neurologic disability in experimental spinal cord ischemic injury. Proc Natl Acad Sci USA 99:2258-2263.

Chavez JC, LaManna JC (2002) Activation of hypoxia-inducible factor-1 in the rat cerebral cortex after transient global ischemia: potential role of insulin-like growth factor-1. J Neurosci 22:8922-8931.

Chong ZZ, Kang JQ, Maiese K (2002) Erythropoietin is a novel vascular protectant through activation of Akt1 and mitochondrial modulation of cysteine proteases. Circulation 106:2973-2979.

Davis JM, Arakawa T, Strickland TW, Yphantis DA (1987) Characterization of recombinant human erythropoietin produced in Chinese hamster ovary cells. Biochemistry 26:2633-2638.

Digicaylioglu M, Bichet S, Marti HH, Wenger RH, Rivas LA, Bauer C, Gassmann M (1995) Localization of specific erythropoietin binding sites in defined areas of the mouse brain. Proc Natl Acad Sci USA 92:3717-3720.

Digicaylioglu M, Garden G, Timberlake S, Fletcher L, Lipton SA (2004) Acute neuroprotective synergy of erythropoietin and insulin-like growth factor I. Proc Natl Acad Sci USA 101:9855-9860.

Dirnagl U, Simon RP, Hallenbeck JM (2003) Ischemic tolerance and endogenous neuroprotection. Trends Neurosci 26:248-254.

Doetsch F (2003) The glial identity of neural stem cells. Nature Neurosci 6:1127-1134.

Ehrenreich H, Siren AL (2001a) Benefits of recombinant human erythropoietin on cognitive function. Erythropoiesis: new dimensions in the treatment of anaemia 11:35-40.

Ehrenreich H, Siren AL (2001b) Neuroprotection - what does it mean? - what means do we have? Eur Arch Psyciatry Clin Neurosci 251:149-151.

Ehrenreich H, Hasselblatt M, Dembowski C, Cepek L, Lewczuk P, Stiefel M, Rustenbeck HH, Breiter N, Jacob S, Knerlich F, Bohn M, Poser W, Rüther E, Kochen M, Gefeller O, Gleiter C, Wessel TC, De Ryck M, Itri L, Prange H, Cerami A, Brines M, Siren AL (2002) Erythropoietin therapy for acute stroke is both safe and beneficial. Mol Med 8:495-505.

Erecinska M, Silver IA (2001) Tissue oxygen tension and brain sensitivity to hypoxia. Respir Physiol 128:263-276.

Genc S, Koroglu TF, Genc K (2004) Erythropoietin and the nervous system. Brain Res 1000:19-31.

Guan J, Bennet L, Gluckman PD, Gunn AJ (2003) Insulin-like growth factor-1 and post-ischemic brain injury. Prog Neurobiol 70:443-462.

Heeschen C, Aicher A, Lehmann R, Fichtlscherer S, Vasa M, Urbich C, Mildner-Rihm C, Martin H, Zeiher AM, Dimmeler S (2003) Erythropoietin is a potent physiologic stimulus for endothelial progenitor cell mobilization. Blood 102:1340-1346.

Juul SE, Harcum J, Li Y, Christensen RD (1997) Erythropoietin is present in the cerebrospinal fluid of neonates. J Pediatr 130:428-430.

Juul S (2002) Erythropoietin in the central nervous system, and its use to prevent hypoxic-ischemic brain damage. Acta Paediatr Suppl 91:36-42.

Juul SE, McPherson RJ, Farrell FX, Jolliffe L, Ness DJ, Gleason CA (2004) Erytropoietin concentrations in cerebrospinal fluid of nonhuman primates and fetal sheep following high-dose recombinant erythropoietin. Biol Neonate 85:138-144.

Kawakami M, Sekiguchi M, Sato K, Kozaki S, Takahashi M (2001) Erythropoietin receptor-mediated inhibition of exocytotic glutamate release confers neuroprotection during chemical ischemia. J Biol Chem 276:39469-39475.

Konishi Y, Chui D-H, Hirose H, Kunishita T, Tabira T (1993) Trophic effect of erythropoietin and other hematopoietic factors on central cholinergic neurons in vitro and in vivo. Brain Res 609:29-35.

Koshimura K, Murakami Y, Sohmiya M, Tanaka J, Kato Y (1999) Effects of erythropoietin on neuronal activity. J Neurochem 72:2565-2572.

Lipton P (1999) Ischemic cell death in brain neurons. Physiol Rev 79:1431-1568.

Liu C, Shen K, Liu ZY, Noguchi CT (1997) Regulated human erythropoietin receptor expression in mouse brain. J Biol Chem 272:32395-32400.

Marti HH, Wenger RH, Rivas LA, Straumann U, Digicaylioglu M, Henn V, Yonekawa Y, Bauer C, Gassmann M (1996) Erythropoietin gene expression in human, monkey and murine brain. Eur J Neurosci 8:666-676.

Marti HH, Gassmann M, Wenger RH, Kvietikova I, Morganti-Kossmann MC, Kossmann T, Trentz O, Bauer C (1997) Detection of erythropoietin in human liquor: Intrinsic erythropoietin production in the brain. Kidney Int 51:416-418.

Marti HH, Bernaudin M, Petit E, Bauer C (2000) Neuroprotection and angiogenesis: A dual role of erythropoietin in brain ischemia. News Physiol Sci 15:225-229.

Marti HH, Bernaudin M (2003) Function of erythropoietin in the brain. In: Erythropoietin: molecular biology and clinical use (Jelkmann W, ed), pp195-215. Johnson City (TN): FP Graham Publishing Co.

Marti HH (2004) Erythropoietin and the hypoxic brain. J Exp Biol 207: 3233-3242.

Masuda S, Okano M, Yamagishi K, Nagao M, Ueda M, Sasaki R (1994) A novel site of erythropoietin production: oxygen-dependent production in cultured rat astrocytes. J Biol Chem 269:19488-19493.

Masuda S, Chikuma M, Sasaki R (1997) Insulin-like growth factors and insulin stimulate erythropoietin production in primary cultured astrocytes. Brain Res 746:63-70.

Miu AC, Olteanu AI, Chis I, Heilman RM (2004) Have no fear, erythropoietin is here: erythropoietin protects fear conditioning performances after functional inactivation of the amygdala. Behav Brain Res: in press.

Morishita E, Masuda S, Nagao M, Yasuda Y, Sasaki R (1997) Erythropoietin receptor is expressed in rat hippocampal and cerebral cortical neurons, and erythropoietin prevents in vitro glutamate-induced neuronal death. Neuroscience 76:105-116.

Nagai A, Nakagawa E, Choi HB, Hatori K, Kobayashi S, Kim SU (2001) Erythropoietin and erythropoietin receptors in human CNS neurons, astrocytes, microglia, and oligodendrocytes grown in culture. J Neuropathol Exp Neurol 60:386-392.

Nitta K, Uchida K, Kimata N, Honda K, Kobayashi H, Kawashima A, Yumura W, Nihei H (1999) Recombinant human erythropoietin stimulates vascular endothelial growth factor release by glomerular endothelial cells. Eur J Pharmacol 373:121-124.

Prass K, Scharff A, Ruscher K, Löwl D, Muselmann C, Victorov I, Kapinya K, Dirnagl U, Meisel A (2003) Hypoxia-induced stroke tolerance in the mouse is mediated by erythropoietin. Stroke 34:1981-1986.

Recny MA, Scoble HA, Kim Y (1987) Structural characterization of natural human urinary and recombinant DNA-derived erythropoietin. Identification of des-arginine 166 erythropoietin. J Biol Chem 262:17156-17163.

Ribatti D, Presta M, Vacca A, Ria R, Giuliani R, Dell'Era P, Nico B, Roncali L, Dammacco F (1999) Human erythropoietin induces a pro-angiogenic phenotype in cultured endothelial cells and stimulates neovascularization in vivo. Blood 93:2627-2636.

Risau W, Flamme I (1995) Vasculogenesis. Annu Rev Cell Dev Biol 11:73-91.

Risau W (1997) Mechanisms of angiogenesis. Nature 386:671-674.

Ruscher K, Freyer D, Karsch M, Isaev N, Megow D, Sawitzki B, Priller J, Dirnagl U, Meisel A (2002) Erythropoietin is a paracrine mediator of ischemic tolerance in the brain: evidence from an in vitro model. J Neurosci 22:10291-10301.

Sadamoto Y, Igase K, Sakanaka M, Sato K, Otsuka H, Sakaki S, Masuda S, Sasaki R (1998) Erythropoietin prevents place navigation disability and cortical infarction in rats with permanent occlusion of the middle cerebral artery. Biochem Biophys Res Commun 253:26-32.

Sakanaka M, Wen TC, Matsuda S, Masuda S, Morishita E, Nagao M, Sasaki R (1998) In vivo evidence that erythropoietin protects neurons from ischemic damage. Proc Natl Acad Sci USA 95:4635-4640.

Schofield CJ, Ratcliffe PJ (2004) Oxygen sensing by HIF hydroxylases. Nature Rev Mol Cell Biol 5:343-354.

Sharp FR, Bernaudin M (2004) HIF1 and oxygen sensing in the brain. Nature Rev Neurosci 5:437-448.

Shingo T, Sorokan ST, Shimazaki T, Weiss S (2001) Erythropoietin regulates the in vitro and in vivo production of neuronal progenitors by mammalian forebrain neural stem cells. J Neurosci 21:9733-9743.

Siren AL, Knerlich F, Poser W, Gleiter CH, Brück W, Ehrenreich H (2001) Erythropoietin and erythropoietin receptor in human ischemic/hypoxic brain. Acta Neuropathol 101:271-276.

Smith KJ, Bleyer AJ, Little WC, Sane DC (2003) The cardiovascular effects of erythropoietin. Cardiovasc Res 59:538-548.

Studer L, Csete M, Lee SH, Kabbani N, Walikonis J, Wold B, McKay R (2000) Enhanced proliferation, survival, and dopaminergic differentiation of CNS precursors in lowered oxygen. J Neurosci 20:7377-7383.

Sugawa M, Sakurai Y, Ishikawa-Ieda Y, Suzuki H, Asou H (2002) Effects of erythropoietin on glial cell development; oligodendrocyte maturation and astrocyte proliferation. Neurosci Res 44:391-403.

Tan CC, Eckardt K-U, Firth JD, Ratcliffe PJ (1992) Feedback modulation of renal and hepatic erythropoietin mRNA in response to graded anemia and hypoxia. Am J Physiol 263:F474-F481.

Tang YP, Shimizu E, Dube GR, Rampon C, Kerchner GA, Zhuo M, Liu G, Tsien JZ (1999) Genetic enhancement of learning and memory in mice. Nature 401:63-69.

Victoria M, Arroyo A, Castilla MA, Pacheco FRG, Tan D, Riesco A, Casado S, Caramelo C (1998) Role of vascular endothelial growth factor on erythropoietin-related endothelial cell proliferation. J Am Soc Nephrol 9:1998-2004.

Villa P, Bigini P, Mennini T, Agnello D, Laragione T, Cagnotto A, Viviani B, Marinovich M, Cerami A, Coleman TR, Brines M, Ghezzi P (2003) Erythropoietin selectively attenuates cytokine production and inflammation in cerebral ischemia by targeting neuronal apoptosis. J Exp Med 198:971-975.

Weber A, Maier RF, Hoffmann U, Grips M, Hoppenz M, Aktas AG, Heinemann U, Obladen M, Schuchmann S (2002) Erythropoietin improves synaptic transmission during and following ischemia in rat hippocampal slice cultures. Brain Res 958:305-311.

Wiesener MS, Jürgensen JS, Rosenberger C, Scholze CK, Hörstrup JH, Warnecke C, Mandriota S, Bechmann I, Frei UA, Pugh CW, Ratcliffe PJ, Bachmann S, Maxwell PH, Eckardt KU (2003) Widespread hypoxia-inducible expression of HIF-2α in distinct cell populations of different organs. FASEB J 17:271-273.

Yamaji R, Okada T, Moriya M, Naito M, Tsuruo T, Miyatake K, Nakano Y (1996) Brain capillary endothelial cells express two forms of erythropoietin receptor mRNA. Eur J Biochem 239:494-500.

Yamamoto M, Koshimura K, Kawaguchi M, Sohmiya M, Murakami Y, Kato Y (2002) Stimulating effect of erythropoietin on the release of dopamine and acetylcholine from the rat brain slice. Neurosci Lett 292:131-133.

Yasuda Y, Masuda S, Chikuma M, Inoue K, Nagao M, Sasaki R (1998) Estrogen-dependent production of erythropoietin in uterus and its implication in uterine angiogenesis. J Biol Chem 273:25381-25387.

Zhang SC (2001) Defining glial cells during CNS development. Nature Rev Neurosci 2:840-843.

Chapter 3

ERYTHROPOIETIN AND NEUROPROTECTION IN THE CENTRAL NERVOUS SYSTEM: INTRACELLULAR SIGNALING PATHWAYS

Murat Digicaylioglu
The Burnham Institute, 10901 North Torrey Pines Rd, La Jolla, California 92037, USA.

Abstract: Over the past few years, there has been tremendous progress in the research of erythropoietin (EPO) mediated neuroprotection. Signaling events have been identified by the meticulous work of many excellent research groups. The aim of this chapter is to identify and discuss the key signaling molecules and events published in numerous reports that are involved in EPO mediated neuroprotection. In order to provide a better overview, the signaling molecules discussed in this chapter have been divided into acute and chronic signaling events. However, the separation of acute and chronic signaling and the transition from one to the other may be less distinct in reality. More than likely, these signaling molecules are involved in either signaling events at one time or another. Better understanding of the intricacies of EPO signaling in the central and peripheral nervous system will further current research and help provide new strategies for novel therapies.

Key words: neurodegeneration, HIV, Akt, tau, PI-3 kinase

1. INTRODUCTION:

Historically, erythropoietin (EPO) was considered to be the principle regulator of red blood cell production. After an expedition team observed that acute exposure to high altitudes increaseded the number of erythrocytes in the blood, it was suggested that erythrocyte production was being regulated by a humoral factor "hemopoietin" (Fandrey, 2004). Later, the oxygen dependent regulation of EPO expression and its essential role in hematopoiesis was established.

EPO is synthesized by the adult kidney (Koury and Bondurant, 1990; Jelkmann, 1992; Koury, 1992) and is an anti-apoptotic cytokine, which enables committed erythroid progenitor cells to survive (Jelkmann, 1992). As we and others have shown, both EPO and its receptor (EPOR) are expressed in the mammalian central nervous system (CNS) (Masuda, 1993; Digicaylioglu et al., 1995; Marti et al., 1996; Morishita, 1997; Chin et al., 2000; Weishaupt et al., 2004). In recent years the neuroprotective role of EPO was reported in numerous publications, indicating that EPO is a multifunctional trophic factor that possibly has tissue specific functions.

As in the adult kidney, hypoxia is the main stimulus for EPO production in the CNS (Digicaylioglu et al., 1995; Marti et al., 1996; Bernaudin et al., 2000). Recent reports indicate that exogenous EPO is neuroprotective in animal models of cerebral hypoxia/ischemia (stroke), neurodegenerative diseases (Sakanaka et al., 1998; Brines et al., 2000; Sinor and Greenberg, 2000; Siren et al., 2001; Kumral et al., 2003; Prass et al., 2003; Solaroglu et al., 2003; Villa et al., 2003), retinal degeneration (Grimm et al., 2002; Junk et al., 2002; Weishaupt et al., 2004), experimental spinal cord injuries (Celik et al., 2002; Gorio et al., 2002) and gp120/HIV (Digicaylioglu et al., 2004a). EPO signaling through the EPOR is also required for the normal development of the brain in mice (Buemi et al., 2002; Yu et al., 2002). In non-neuronal cells, ligand binding to the EPOR is known to induce activation of the Janus family of protein tyrosine kinase-2 (JAK2) and nuclear translocation of the signal transducer and activator of transcription-5 (STAT5) (Yousouffian, 1993; Ihle and Kerr, 1995; Kirito et al., 1997; Oda et al., 1998; Verdier et al., 1998; Uddin et al., 2000; Gorio et al., 2002; Yu et al., 2002). EPO binding to its receptor also induces complex formation of JAK2 with the phosphorylated EPOR and activation of PI-3 kinase (Damen et al., 1993; Mayeux et al., 1993; Klingmuller et al., 1997; Sui et al., 1998; Weishaupt et al., 2004).

It is well established that EPO is also involved in the control of caspases in neurons and upregulates both the XIAP and Bcl-2 families of anti-apoptotic proteins (Gregoli and Bondurant, 1999; Digicaylioglu and Lipton, 2001). In erythroid progenitor cells, EPO stimulates phosphorylation of Bcl-2 and prevents apoptosis (Ito et al., 1997) while EPO deprivation results in enhanced activation of caspase-3 and cell death (Gregoli and Bondurant, 1999). In addition to its direct protective function on neurons, EPO has a dual role and provides indirect neuroprotection by promoting angiogenesis through inducing mitogenesis of endothelial cells (Yasuda, 1993; Marti et al., 2000) and modulating the astroglial antioxidant defense system (Genc et al., 2001; Genc et al., 2002) by stimulating gluthatione-peroxidase, thereby reducing levels of neurotoxic reactive oxygen species.

The focus of the present chapter is the intracellular signaling involved in the chronic neuroprotection provided by EPO and acute neuroprotection by EPO and insulin-like growth factor (IGF-I) in the CNS.

2. ACUTE AND CHRONIC NEUROPROTECTIVE SIGNALING

In cultured rat cerebrocortical neurons, binding of EPO initiates the activation of EPOR (Digicaylioglu and Lipton, 2001; Chong et al., 2002). Activated EPOR then initiates the activation of numerous signaling molecules in neurons, which are discussed below.

2.1 Signaling pathways for chronic neuroprotection:

EPO is neuroprotective in vitro when its application precedes the onset of the experimental neuronal injury. In our experiments a 3h preincubation with EPO afforded the maximum protection against NMDA induced neurotoxicity (Digicaylioglu and Lipton, 2001).

2.1.1 JAK2 and STAT's:

In non-neuronal cells, JAK2 association with EPOR plays a crucial role in triggering the downstream responses in neurons. JAK2 phosphorylates intracellular tyrosine residues of the EPOR, which are thought to provide docking sites for intracellular signaling molecules (Ihle and Kerr, 1995). However, in animals lacking these residues no difference in erythropoiesis was observed (Zhang, 2001). Best known among the intracellular signaling molecules are STAT 1, 3, 5a and 5b (Schindler, 1999; Schindler and Strehlow, 2000). STAT proteins are constitutively expressed and present in the cytosol of neuronal cells and recruited to the phosphorylated tyrosine residues Y343, Y401 of the activated EPOR complex (Damen et al., 1995a; Damen et al., 1995b; Quelle et al., 1996). JAK phosphorylation of STAT's results in their dimerization and translocation into the nucleus, where they bind to specific sequences in the promoter of STAT-regulated genes (Ihle and Kerr, 1995). Although disruption of functionality of STAT 5a and b results in higher levels of apoptosis in EPO-dependent cells (Socolovsky et al., 1999), in neurons the participation of STAT5 in EPO signaling is disputed. In our experiments we have not observed any loss in EPO mediated protection in neurons expressing the dominant negative non-phosphorylatable form of STAT5a/b (Digicaylioglu and Lipton, 2001). We concluded that STAT5 is not involved in EPO mediated chronic neuroprotection. Supporting our results, Ruscher et al (Ruscher et al., 2002) could not detect any functional role for STAT in hippocampal neurons under ischemic conditions, although JAK2 activation was essential for EPO signaling. However, other groups have reported activation of neural STAT5 by JAK2 (De-Fraja et al., 1998; Chong et al., 2002). Moreover, Bittorf et al (Bittorf et al., 2001) showed that in presence of EPO cell lines expressing the truncated form of STAT5 with sustained binding to specific DNA sequences underwent less apoptosis than cells expressing the wild-type STAT5 (Bittorf et al., 2001). Interestingly, one report indicates that STAT5 is activated in axotomized neurons in the peripheral nervous system but not in CNS (Schwaiger et al., 2000; Liu and Snider, 2001) present some evidence that STAT5 might be a downstream target of JAK2 in regeneration but not in development of peripheral sensory neurons. It is conceivable that different signaling molecules are involved in EPO mediated chronic neuroprotection and in development or regeneration. Furthermore, downstream targets of EPO-activated JAK2 might be cell-specific and not always utilize STAT5.

2.1.2 IKK/ NF-κB /IκB:

The NF-κBs are a family of transcription factors that have been implicated to play a role in survival and apoptotic signaling pathways. These factors are sequestered in the cytoplasm by IκBs, which lose their ability to bind NF-κB when their ubiquitin-dependent degradation is initiated via phosphorylation, usually by IKK. NF-κB is then free to translocate to the nucleus and bind to DNA. NF-κB targets a number of genes that are both pro- and anti-apoptotic, including p53, c-myc, Fas, Bcl-x, Bcl-2, XIAP, cIAP2, and MnSOD (for review, see Karin et al, 2003).

Classically, NF-κB is known to operate in immune cells, mediating the inflammatory response, but was later found to be expressed continuously in both neurons and glia (Kaltschmidt et al 1994). Activation of NF-κB in glial cells leads to expression of inflammatory proteins that cause apoptosis of neurons in mixed cultures. The role of NF-κB in neurons is complex; both pro-and anti-apoptotic functions have been described in a number of varied treatments and models. Several lines of evidence support the hypothesis that acute increases in NF-κB activate an apoptotic signaling pathway, whereas stimuli that lead to large increases in steady-state NF-κB activity provide neuroprotection (Lin et al 1998, Shen et al 2002, Aleyasin et al 2004). It is likely that the cellular responses to NF-κB vary with the specific environment of the cell—its current transcriptome.

NF-κB is predominantly localized in the cytoplasm of neuronal cells under basal conditions (Digicaylioglu 2001) but a variety of treatments and conditions trigger its activation. Pro-apoptotic roles for NF-κB in neuronal cells in vivo have been described in several animal models. Excitotoxicity resulted in increased nuclear translocation of NF-κB and consequent upregulation of apoptotic genes and cell death (Qin et al 1999), but these effects were abrogated by interfering with translocation of NF-κB to the nucleus. Ischemia in a transgenic model provided additional support for the pro-apoptotic role of NF-κB in acute cellular trauma (Schnieder et al 1999). However, similar methods have discerned protective roles for the protein in neurons (Botchkina et al, 1999).

In vitro studies have also found both protective and degenerative roles for NF-κB (for review, see Barkett and Gilmore, 1999) in the nervous system. A basal rate of activity is critical for the survival of primary cortical neurons in culture but numerous in vitro insults lead to acute activation and subsequent apoptosis (reviewed in Barkett and Gilmore, 1999). In the neuronal nucleus, active NF-κB induces a number of pro- and anti-apoptotic proteins. Pharmacologic, functional, and genetic inhibitors of NF-κB can increase neuronal death upon prolonged exposure (Aleyasin et al 2004, Natarajan et al, 1996). However, these same methods of inhibition can provide protection to cultured neurons against death from acute insults, such as excitotoxicity and DNA damage. (Rothman et al 1986, Grilli et al 1996, Aleyasin et al 2004).

High levels of constitutively active NF-κB are known to mediate the extraordinary resistance of some neuronal cell lines to oxidative stress (Lezoualc'h et al, 1998). Because EPO is capable of affecting chronic neuroprotection, we investigated whether it was able to act through NF-κB to protect neurons from excitotoxic and nitrosative stress (Digicaylioglu 2001). Treatment of cerebrocortical cultures with EPO resulted in nuclear translocation, and a large, sustained increase in DNA-binding activity of NF-κB in neurons (but not astrocytes). Expression of a reporter gene confirmed the transcriptional activity of NF-κB in this model. EPO treatment caused JAK2 to become phosphorylated, and, in turn, to directly phosphorylate IκBκ. Inhibition of this pathway, either by pharmacological inhibition of JAK2, or expression of a dominant interfering form of JAK2 or a IκB super-repressor construct, significantly attenuated NF-κB DNA-binding and reporter gene transcription in response to EPO, and abrogated EPO-mediated neuroprotection. The effects of interrupting this signaling pathway thus confirm its importance in sustained anti-apoptotic signaling initiated by EPO treatment.

Supplementing this in vitro evidence for the protection conferred by chronic NF-κB activation, an in vivo study demonstrated its importance in attenuating the neurological damage of stroke. Sub-lethal insults are known in vivo to render neurons resistant to subsequent, more serious episodes, via a process known as tolerance, or pre-conditioning (Kitagawa et al, 1990). A short bout of ischemia was found to trigger long-term activation of NF-κB, the key event in pre-conditioning the brain against later, severe episodes of ischemia (Blondeau et al, 2001). So despite the pro-apoptotic role of NF-κB in neuronal signal transduction of acute insults, chronic increases of NF-κB activity (such as those triggered by treatment with EPO) have been shown, both in vivo and in vitro, to protect against a number of neurotoxic treatments.

2.1.3 XIAP and caspase activity

The inhibitors-of-apoptosis (IAP) molecules are factors known to prevent neuronal cell death under a variety of conditions (Deveraux and Reed, 1999). Among these factors is X-linked IAP (XIAP), which prevents neurodegeneration (Holcik and Korneluk, 2001). The importance of XIAP in preventing of neuronal degeneration is outlined in the transgenic animals model overexpressing XIAP (Trapp et al., 2003). In the transgenic mouse, overexpression of XIAP in neurons (Trapp et al., 2003) resulted in significant neuroprotection from transient cerebral ischemia. We have shown, that EPO causes XIAP is upregulation and facilitates its binding to activated caspase-3 (Digicaylioglu and Lipton, 2001), resulting in reduced neuronal death.

2.2 Signaling pathways involved in the acute neuroprotection

A required preincubation for 3-8 h (Morishita, 1997; Digicaylioglu and Lipton, 2001) before the onset of a neurotoxic insult limits the potential for EPO to be used as an acute neuroprotective treatment. We have found that co-incubating EPO with IGF-I results in acute and prolonged neuroprotection *in vitro* (Digicaylioglu et al., 2004b). Therefore we will discuss the pathways activated by the combination of EPO and IGF-I.

2.2.1 JAK2

As described for the EPO mediated chronic neuroprotection, JAK2 signaling is also involved in acute neuroprotection. Previously we reported the association of JAK2 with the EPOR complex in cortical neurons (Digicaylioglu and Lipton, 2001). Association of JAK2 and the EPOR may lead to phosphorylation and activation of phosphatidylinositol-3 kinases (PI-3K) (Pleiman et al., 1994; Scheid and Woodgett, 2001). Although direct functional evidence is lacking, ligand binding to the neuronal EPOR promotes the association of p85 and JAK2 with the EPOR (Digicaylioglu et al., 2004b), which might facilitate the activation of signaling pathways induced by IGF-I.

2.2.2 STAT's

In non-neuronal EPO/EPOR signaling STAT5 is required. In embryonic $STAT^{-/-}$ mice severe anemia is caused by the disruption of the EPO/JAK2/STAT5 signaling pathway (Socolovsky et al., 1999). However, it would be interesting to see if the brains of the $STAT^{-/-}$ mice show any abnormal development. Currently we have no evidence from our work that STAT's are involved either in chronic or acute neuroprotective signaling by EPO. As described above for chronic neuroprotection, other groups have reported the involvement of STAT5 or other STAT's in EPO/JAK2 signaling in neurons. However, none of the published studies used a specific blocker for STAT's, such as a dominant-negative form of STAT or siRNA. This would enable us to determine if the EPO signaling is disrupted in absence of functional STAT proteins. It is possible that activation of STAT's in these reports is a secondary effect, not directly related to neuroprotective signaling.

2.2.3 Phosphatidylinositol-3 kinases (PI-3 K)

PI-3Ks are members of a lipid kinase family (Fry and Waterfield, 1993). PI-3Ks consist of the catalytic subunit p110 and the regulatory subunit p85, and are activated by receptor protein tyrosine kinases (RPTKs). PI-3Ks phosphorylate and activate the inositol 3'-OH group in inositol phospholipids, resulting in the second messengers phosphatidylinositol-3,4,5-trisphosphate (PI-3,4,5-P_3) and -diphosphate (PI-3,4-P_2). Interaction with these phospholipids results in activation of Akt.

In non-neuronal cells EPO-mediated phosphorylation and activation of the PI-3Ks has been reported extensively (LeRoith and Roberts, 1993; Mayeux et al., 1993; Pleiman et al., 1994; Klingmuller et al., 1997; Park et al., 2001). Activated EPOR provides binding sites for the regulatory subunit of PI-3K, p85 (Nguyen et al., 2001; Digicaylioglu et al., 2004b). Phosphorylation and subsequent activation of p85 results in association of its Src2 homology domains (SH2) with EPOR (Damen et al., 1993) and the generation of second messengers. Subsequent activation of Akt by PI-3K is required for neuroprotection (Ruscher et al., 2002; Chong et al., 2003; Weishaupt et al., 2004). Using a pharmacological inhibitor, we demonstrated that PI-3K is required for the cooperative and acute neuroprotective effects of EPO+IGF-I (Digicaylioglu et al., 2004b). Our results suggest that concurrent activation of PI-3K by IGF-I enables EPO to act as an acute neuroprotectant, independent of the more prolonged effects on gene expression mediated by the JAK2/NF-κB pathway that we have previously demonstrated (Digicaylioglu and Lipton, 2001).

2.2.4 Akt

In acute neuroprotection, Akt plays a crucial role in mediating EPO/JAK2/PI-3K signaling (Digicaylioglu et al., 2004b). As a serine-threonine kinase, activated Akt phosphorylates its downstream targets, among them Bad (Li et al 2001), caspase-9 (Zhou et al., 2000; Kaspar et al., 2003; Manabe et al., 2003), Forkhead transcription factor (Brunet et al, 1999) and glycogen synthase kinase 3beta (GSK-3β, Cross et al 1995), thereby decreasing or blocking neuronal apoptosis (Noshita et al., 2002). Cysteine aspartases, called caspases, are activated during neuronal apoptosis and lead to cell death (Green and Reed, 1998; Krajewski et al., 1999; Nicholson, 1999). Activation of the "initiator" caspase-9 (Krajewski et al., 1999) leads to subsequent activation of the "executioner" caspase-3 (Okamoto et al., 2002; Bossy-Wetzel et al., 2004). Inhibition of caspase-3 activity by a specific blocker, Z-DEVD-FMK, attenuates neuronal death (Endres et al., 1998). Under apoptotic conditions, caspase-9 activates caspase-3, which directly induces DNA fragmentation by activating DNases. Akt blocks caspase-3 activation by phosphorylating the proapoptotic molecule BAD, which suppresses expression and activation of the antiapoptotic factor Bcl-X_L. In addition, the enzymatic activity of caspase-3 can be inhibited by binding to XIAP (see below), offering an additional protective potential. Using a dominant interfering mutant of Akt (dn-Akt), we demonstrated that Akt is required for the cooperative and acute neuroprotective effects of EPO+IGF-I. Moreover, our preliminary results indicate that EPO+IGF-I combination greatly reduces Akt dephosphorylation (Digicaylioglu et al., 2004b).

2.2.5 GSK3-β and Tau proteins

GSK3-β, another downstream target of Akt, promotes apoptosis in neurons (Crowder and Freeman, 2000; Hetman et al., 2000; Mora et al., 2001; Higuchi et al., 2003; Jones et al., 2003) possibly through hyperphosphorylation of tau and beta-catenin (Lucas et al., 2001; Schubert et al., 2004). Therefore Akt hyperphosphorylation and inhibition of pro-apoptotic GSK-3β contributes to EPO+IGF-I mediated neuroprotection. In agreement with this, overexpression of GSK-3β results in increased neuronal apoptosis (Bhat et al., 2000). Similarly, expression of non-inducible form of

Akt, which lacks serine phosphorylation ability (Fujio and Walsh, 1999) results in activation of GSK-3β by de-phophorylation and higher neuronal apoptosis (Crowder and Freeman, 2000; Noshita et al., 2002; Stoica et al., 2003).

2.2.6 XIAP and caspase activity

Our results indicate that as an acute neuroprotectant, EPO in combination with IGF-I induces binding of XIAP to active caspase-3 and inhibits the proteolytic activity of the "executioner" caspases downstream of caspase-3 activation (Digicaylioglu et al., 2004b). As a chronic neuroprotectant, EPO also induces and regulates XIAP and cIAP-2 expression in neurons (Digicaylioglu and Lipton, 2001). Acute increase in XIAP and cIAP-2 expression and reduction in caspase activity for chronic protection by increased association of the inhibitors of apoptosis with active caspase-3 and subsequent block of "executioner" caspases downstream affords more sustained neuroprotection.

2.2.7 IKK, IκBα, NF-κB

In the initial phase of acute EPO+IGF-I neuroprotection, NF-κB signaling seems to uninvolved. However, as a key signaling event in chronic neuroprotection, NF-κB must be activated during the transition from acute to chronic neuroprotection. Akt is known to phosphorylate and activate IKK, increasing, resulting in dissociation of IκB from the p65/p50 heterodimer (Romashkova and Makarov, 1999) and enhanced NF-κB function in non-neuronal cells (Ozes et al., 1999; Burow et al., 2000). Akt can also influence NF-κB directly, through activation of MAP3K (Li et al 1998). The effect of PI-3K on NF-κB activity seems to be mediated primarily by IKK, and inhibitors of PI-3K block NF-κB DNA binding in IKK-/- but not IKK-/- cells (Gustin et al, 2004).

However, in neuronal cells there is only little evidence for the participation of PI3K/Akt pathway in NF-κB activation (Min et al., 2003) during EPO+IGF-I induced acute neuroprotection. In a different experimental model Bittorf et al (Bittorf et al., 2001) confirmed the requirement of NF-κB activation for EPO mediated neuroprotection, although this signaling pathway was independent of JAK2, but required activation of src-kinases.

3. CONCLUSION AND FUTURE DIRECTIONS

As evidenced by the number of publications and the resulting insight, EPO emerges as a novel neuroprotectant with significant clinical potential. However, novel application methods in humans have to be investigated to prevent undesired side effects of systemically applied EPO on hematocrit levels. This is particularly important in chronic treatment with EPO. Recently, an engineered molecule similar to EPO, carbamylated EPO (CEPO), was shown to have neuroprotective effects comparable to recombinant human EPO (Leist et al., 2004). However, this molecule does not bind to EPOR homodimers or exert a hematopoietic effect. Moreover, none of the known signaling pathways, such as JAK2 and STAT5, are activated by CEPO at concentrations sufficient to protect neurons. Therefore, it remains to be determined whether this molecule has any relation to EPO.

Detailed knowledge about EPO signaling in neurons is still scarce and more studies about the neuronal signaling pathways involved in EPO mediated protection are required. In particular, the transition in intracellular signaling from the acute to the chronic effect has to be more elucidated, in order to use EPO alone or in combination with other cytokines as a therapeutic molecule in humans. The combined use of EPO and IGF-I can provide a powerful tool for the acute therapy of stroke and CNS trauma. EPO should also be considered for the chronic therapy and the management of neurodegenerative diseases, such as autism, ALS, Alzheimer's disease, Parkinson's disease and AIDS dementia. In addition, the potential for EPO as a pediatric therapeutic for neurodevelopmental diseases is still widely unexplored and deserves more attention.

REFERENCES:

Bernaudin M, Bellail A, Marti HH, Yvon A, Vivien D, Duchatelle I, Mackenzie ET, Petit E (2000) Neurons and astrocytes express EPO mRNA: Oxygen-sensing mechanisms that involve the redox-state of the brain. Glia 30:271-278.

Bhat RV, Shanley J, Correll MP, Fieles WE, Keith RA, Scott CW, Lee CM (2000) Regulation and localization of tyrosine216 phosphorylation of glycogen synthase kinase-3beta in cellular and animal models of neuronal degeneration. Proc Natl Acad Sci U S A 97:11074-11079.

Bittorf T, Buchse T, Sasse T, Jaster R, Brock J (2001) Activation of the transcription factor NF-kappaB by the erythropoietin receptor. Structural requirements and biological significance. Cell Signal 13:673-681.

Bossy-Wetzel E, Talantova MV, Lee WD, Scholzke MN, Harrop A, Mathews E, Gotz T, Han J, Ellisman MH, Perkins GA, Lipton SA (2004) Crosstalk between nitric oxide and zinc pathways to neuronal cell death involving mitochondrial dysfunction and p38-activated K+ channels. Neuron 41:351-365.

Brines ML, Ghezzi P, Keenan S, Agnello D, de Lanerolle NC, Cerami C, Itri LM, Cerami A (2000) Erythropoietin crosses the blood-brain barrier to protect against experimental brain injury. Proc Natl Acad Sci U S A 97:10526-10531.

Buemi M, Cavallaro E, Floccari F, Sturiale A, Aloisi C, Trimarchi M, Grasso G, Corica F, Frisina N (2002) Erythropoietin and the brain: from neurodevelopment to neuroprotection. Clin Sci (Lond) 103:275-282.

Burow ME, Weldon CB, Melnik LI, Duong BN, Collins-Burow BM, Beckman BS, McLachlan JA (2000) PI3-K/AKT regulation of NF-kappaB signaling events in suppression of TNF-induced apoptosis. Biochem Biophys Res Commun 271:342-345.

Celik M, Gokmen N, Erbayraktar S, Akhisaroglu M, Konakc S, Ulukus C, Genc S, Genc K, Sagiroglu E, Cerami A, Brines M (2002) Erythropoietin prevents motor neuron apoptosis and neurologic disability in experimental spinal cord ischemic injury. Proc Natl Acad Sci U S A 99:2258-2263.

Chin K, Yu X, Beleslin-Cokic B, Liu C, Shen K, Mohrenweiser HW, Noguchi CT (2000) Production and processing of erythropoietin receptor transcripts in brain. Brain Res Mol Brain Res 81:29-42.

Chong ZZ, Kang JQ, Maiese K (2002) Hematopoietic factor erythropoietin fosters neuroprotection through novel signal transduction cascades. J Cereb Blood Flow Metab 22:503-514.

Chong ZZ, Kang JQ, Maiese K (2003) Erythropoietin fosters both intrinsic and extrinsic neuronal protection through modulation of microglia, Akt1, Bad, and caspase-mediated pathways. Br J Pharmacol 138:1107-1118.

Crowder RJ, Freeman RS (2000) Glycogen synthase kinase-3 beta activity is critical for neuronal death caused by inhibiting phosphatidylinositol 3-kinase or Akt but not for death caused by nerve growth factor withdrawal. J Biol Chem 275:34266-34271.

Damen JE, Mui AL, Puil L, Pawson T, Krystal G (1993) Phosphatidylinositol 3-kinase associates, via its Src homology 2 domains, with the activated erythropoietin receptor. Blood 81:3204-3210.

Damen JE, Cutler RL, Jiao H, Yi T, Krystal G (1995a) Phosphorylation of tyrosine 503 in the erythropoietin receptor (EpR) is essential for binding the P85 subunit of phosphatidylinositol (PI) 3- kinase and for EpR-associated PI 3-kinase activity. J Biol Chem 270:23402-23408.

Damen JE, Wakao H, Miyajima A, Krosl J, Humphries RK, Cutler RL, Krystal G (1995b) Tyrosine 343 in the erythropoietin receptor positively regulates erythropoietin-induced cell proliferation and Stat5 activation. Embo J 14:5557-5568.

De-Fraja C, Conti L, Magrassi L, Govoni S, Cattaneo E (1998) Members of the JAK/STAT proteins are expressed and regulated during development in the mammalian forebrain. J Neurosci Res 54:320-330.

Deveraux QL, Reed JC (1999) IAP family proteins--suppressors of apoptosis. Genes Dev 13:239-252.

Digicaylioglu M, Lipton SA (2001) Erythropoietin-mediated neuroprotection involves cross-talk between Jak2 and NF-kappaB signalling cascades. Nature 412:641-647.

Digicaylioglu M, Kaul M, Fletcher L, Dowen R, Lipton SA (2004a) Erythropoietin protects cerebrocortical neurons from HIV-1/gp120-induced damage. Neuroreport 15:761-763.

Digicaylioglu M, Garden G, Timberlake S, Fletcher L, Lipton SA (2004b) Acute neuroprotective synergy of erythropoietin and insulin-like growth factor I. Proc Natl Acad Sci U S A 101:9855-9860.

Digicaylioglu M, Bichet S, Marti HH, Wenger RH, Rivas LA, Bauer C, Gassmann M (1995) Localization of specific erythropoietin binding sites in defined areas of the mouse brain. Proc Natl Acad Sci USA 92:3717-3720.

Endres M, Namura S, Shimizu-Sasamata M, Waeber C, Zhang L, Gomez-Isla T, Hyman BT, Moskowitz MA (1998) Attenuation of delayed neuronal death after mild focal ischemia in mice by inhibition of the caspase family. J Cereb Blood Flow Metab 18:238-247.

Fandrey J (2004) Oxygen-dependent and tissue-specific regulation of erythropoietin gene expression. Am J Physiol Regul Integr Comp Physiol 286:R977-988.

Fry MJ, Waterfield MD (1993) Structure and function of phosphatidylinositol 3-kinase: a potential second messenger system involved in growth control. Philos Trans R Soc Lond B Biol Sci 340:337-344.

Fujio Y, Walsh K (1999) Akt mediates cytoprotection of endothelial cells by vascular endothelial growth factor in an anchorage-dependent manner. J Biol Chem 274:16349-16354.

Genc S, Akhisaroglu M, Kuralay F, Genc K (2002) Erythropoietin restores glutathione peroxidase activity in 1-methyl-4-phenyl-1,2,5,6-tetrahydropyridine-induced neurotoxicity in C57BL mice and stimulates murine astroglial glutathione peroxidase production in vitro. Neurosci Lett 321:73-76.

Genc S, Kuralay F, Genc K, Akhisaroglu M, Fadiloglu S, Yorukoglu K, Fadilo, gbreve, lu M, Gure A (2001) Erythropoietin exerts neuroprotection in 1-methyl-4-phenyl-1,2,3,6-tetrahydropyridine-treated C57/BL mice via increasing nitric oxide production. Neurosci Lett 298:139-141.

Gorio A, Gokmen N, Erbayraktar S, Yilmaz O, Madaschi L, Cichetti C, Di Giulio AM, Vardar E, Cerami A, Brines M (2002) Recombinant human erythropoietin counteracts secondary injury and markedly enhances neurological recovery from experimental spinal cord trauma. Proc Natl Acad Sci U S A 99:9450-9455.

Green DR, Reed JC (1998) Mitochondria and apoptosis. Science 281:1309-1312.

Gregoli PA, Bondurant MC (1999) Function of caspases in regulating apoptosis caused by erythropoietin deprivation in erythroid progenitors. J Cell Physiol 178:133-143.

Grimm C, Wenzel A, Groszer M, Mayser H, Seeliger M, Samardzija M, Bauer C, Gassmann M, Reme CE (2002) HIF-1-induced erythropoietin in the hypoxic retina protects against light-induced retinal degeneration. Nat Med 8:718-724.

Hetman M, Cavanaugh JE, Kimelman D, Xia Z (2000) Role of glycogen synthase kinase-3beta in neuronal apoptosis induced by trophic withdrawal. J Neurosci 20:2567-2574.

Higuchi M, Onishi K, Masuyama N, Gotoh Y (2003) The phosphatidylinositol-3 kinase (PI3K)-Akt pathway suppresses neurite branch formation in NGF-treated PC12 cells. Genes Cells 8:657-669.

Holcik M, Korneluk RG (2001) XIAP, the guardian angel. Nat Rev Mol Cell Biol 2:550-556.

Ihle JN, Kerr IM (1995) Jaks and Stats in signaling by the cytokine receptor superfamily. Trends in Genetics 11:69-74.

Ito T, Deng X, Carr B, May WS (1997) Bcl-2 phosphorylation required for anti-apoptosis function. J Biol Chem 272:11671-11673.

Jelkmann W (1992) Erythropoietin: structure, control of production, and function. Physiological Reviews 72:449-489.

Jones DM, Tucker BA, Rahimtula M, Mearow KM (2003) The synergistic effects of NGF and IGF-1 on neurite growth in adult sensory neurons: convergence on the PI 3-kinase signaling pathway. J Neurochem 86:1116-1128.

Junk AK, Mammis A, Savitz SI, Singh M, Roth S, Malhotra S, Rosenbaum PS, Cerami A, Brines M, Rosenbaum DM (2002) Erythropoietin administration protects retinal neurons from acute ischemia-reperfusion injury. Proc Natl Acad Sci U S A 99:10659-10664.

Kaspar BK, Llado J, Sherkat N, Rothstein JD, Gage FH (2003) Retrograde viral delivery of IGF-1 prolongs survival in a mouse ALS model. Science 301:839-842.

Kirito K, Uchida M, Yamada M, Miura Y, Komatsu N (1997) A distinct function of STAT proteins in erythropoietin signal transduction. J Biol Chem 272:16507-16513.

Klingmuller U, Wu H, Hsiao JG, Toker A, Duckworth BC, Cantley LC, Lodish HF (1997) Identification of a novel pathway important for proliferation and differentiation of primary erythroid progenitors. Proc Natl Acad Sci U S A 94:3016-3021.

Koury MJ (1992) Programmed cell death (apoptosis) in hematopoiesis. Experimental Hematology 20:391-394.

Koury MJ, Bondurant MC (1990) Control of red cell production: the roles of programmed cell death (apoptosis) and erythropoietin. Transfusion 30:673-674.

Krajewski S, Krajewska M, Ellerby LM, Welsh K, Xie Z, Deveraux QL, Salvesen GS, Bredesen DE, Rosenthal RE, Fiskum G, Reed JC (1999) Release of caspase-9 from mitochondria during neuronal apoptosis and cerebral ischemia. Proc Natl Acad Sci U S A 96:5752-5757.

Kumral A, Ozer E, Yilmaz O, Akhisaroglu M, Gokmen N, Duman N, Ulukus C, Genc S, Ozkan H (2003) Neuroprotective effect of erythropoietin on hypoxic-ischemic brain injury in neonatal rats. Biol Neonate 83:224-228.

Leist M, Ghezzi P, Grasso G, Bianchi R, Villa P, Fratelli M, Savino C, Bianchi M, Nielsen J, Gerwien J, Kallunki P, Larsen AK, Helboe L, Christensen S, Pedersen LO, Nielsen M, Torup L, Sager T, Sfacteria A, Erbayraktar S, Erbayraktar Z, Gokmen N, Yilmaz O, Cerami-Hand C, Xie QW, Coleman T, Cerami A, Brines M (2004) Derivatives of erythropoietin that are tissue protective but not erythropoietic. Science 305:239-242.

LeRoith D, Roberts CT, Jr. (1993) Insulin-like growth factors. Ann N Y Acad Sci 692:1-9.

Liu RY, Snider WD (2001) Different signaling pathways mediate regenerative versus developmental sensory axon growth. J Neurosci 21:RC164.

Lucas JJ, Hernandez F, Gomez-Ramos P, Moran MA, Hen R, Avila J (2001) Decreased nuclear beta-catenin, tau hyperphosphorylation and neurodegeneration in GSK-3beta conditional transgenic mice. Embo J 20:27-39.

Manabe Y, Nagano I, Gazi MS, Murakami T, Shiote M, Shoji M, Kitagawa H, Abe K (2003) Glial cell line-derived neurotrophic factor protein prevents motor neuron loss of transgenic model mice for amyotrophic lateral sclerosis. Neurol Res 25:195-200.

Marti HH, Bernaudin M, Petit E, Bauer C (2000) Neuroprotection and Angiogenesis: Dual Role of Erythropoietin in Brain Ischemia. News Physiol Sci 15:225-229.

Marti HH, Wenger RH, Rivas LA, Straumann U, Digicaylioglu M, Henn V, Yonekawa Y, Bauer C, Gassmann M (1996) Erythropoietin gene expression in human, monkey and murine brain. Eur J Neurosci 8:666-676.

Masuda S, Nagao, M., Takahata, K., Konishi, Y., Gallyas, Jr. F., Tabira, T., Sasaki, R. (1993) Functional erythropoietin receptors of the cells with neuronal characteristics-comparison with receptor properties from erythroid cells. J Biol Chem 268:11208-11216.

Mayeux P, Dusanter-Fourt I, Muller O, Mauduit P, Sabbah M, Druker B, Vainchenker W, Fischer S, Lacombe C, Gisselbrecht S (1993) Erythropoietin induces the association of phosphatidylinositol 3'- kinase with a tyrosine-phosphorylated protein complex containing the erythropoietin receptor. Eur J Biochem 216:821-828.

Min YK, Park JH, Chong SA, Kim YS, Ahn YS, Seo JT, Bae YS, Chung KC (2003) Pyrrolidine dithiocarbamate-induced neuronal cell death is mediated by Akt, casein kinase

2, c-Jun N-terminal kinase, and IkappaB kinase in embryonic hippocampal progenitor cells. J Neurosci Res 71:689-700.

Mora A, Sabio G, Gonzalez-Polo RA, Cuenda A, Alessi DR, Alonso JC, Fuentes JM, Soler G, Centeno F (2001) Lithium inhibits caspase 3 activation and dephosphorylation of PKB and GSK3 induced by K+ deprivation in cerebellar granule cells. J Neurochem 78:199-206.

Morishita E, Masuda, S., Nagao, M., Sasaki, R. (1997) Erythropoietin receptor is expressed in rat hippocampal cerebral cortical neurons, and erythropoietin prevents in vitro glutamate-induced neuronal death. Neuroscience 76:105-116.

Nguyen MH, Ho JM, Beattie BK, Barber DL (2001) TEL-JAK2 mediates constitutive activation of the phosphatidylinositol 3'-kinase/protein kinase B signaling pathway. J Biol Chem 276:32704-32713.

Nicholson DW (1999) Caspase structure, proteolytic substrates, and function during apoptotic cell death. Cell Death Differ 6:1028-1042.

Noshita N, Lewen A, Sugawara T, Chan PH (2002) Akt phosphorylation and neuronal survival after traumatic brain injury in mice. Neurobiol Dis 9:294-304.

Oda A, Sawada K, Druker BJ, Ozaki K, Takano H, Koizumi K, Fukada Y, Handa M, Koike T, Ikeda Y (1998) Erythropoietin induces tyrosine phosphorylation of jak2, STAT5A, and STAT5B in primary cultured human erythroid precursors. Blood 92:443-451.

Okamoto S, Li Z, Ju C, Scholzke MN, Mathews E, Cui J, Salvesen GS, Bossy-Wetzel E, Lipton SA (2002) Dominant-interfering forms of MEF2 generated by caspase cleavage contribute to NMDA-induced neuronal apoptosis. Proc Natl Acad Sci U S A 99:3974-3979.

Ozes ON, Mayo LD, Gustin JA, Pfeffer SR, Pfeffer LM, Donner DB (1999) NF-kappaB activation by tumour necrosis factor requires the Akt serine-threonine kinase. Nature 401:82-85.

Park J, Hill MM, Hess D, Brazil DP, Hofsteenge J, Hemmings BA (2001) Identification of tyrosine phosphorylation sites on 3-phosphoinositide-dependent protein kinase-1 and their role in regulating kinase activity. J Biol Chem 276:37459-37471.

Pleiman CM, Hertz WM, Cambier JC (1994) Activation of phosphatidylinositol-3' kinase by Src-family kinase SH3 binding to the p85 subunit. Science 263:1609-1612.

Prass K, Scharff A, Ruscher K, Lowl D, Muselmann C, Victorov I, Kapinya K, Dirnagl U, Meisel A (2003) Hypoxia-induced stroke tolerance in the mouse is mediated by erythropoietin. Stroke 34:1981-1986.

Quelle FW, Wang D, Nosaka T, Thierfelder WE, Stravopodis D, Weinstein Y, Ihle JN (1996) Erythropoietin induces activation of Stat5 through association with specific tyrosines on the receptor that are not required for a mitogenic response. Mol Cell Biol 16:1622-1631.

Romashkova JA, Makarov SS (1999) NF-kappaB is a target of AKT in anti-apoptotic PDGF signalling. Nature 401:86-90.

Ruscher K, Freyer D, Karsch M, Isaev N, Megow D, Sawitzki B, Priller J, Dirnagl U, Meisel A (2002) Erythropoietin is a paracrine mediator of ischemic tolerance in the brain: evidence from an in vitro model. J Neurosci 22:10291-10301.

Sakanaka M, Wen TC, Matsuda S, Masuda S, Morishita E, Nagao M, Sasaki R (1998) In vivo evidence that erythropoietin protects neurons from ischemic damage. Proc Natl Acad Sci U S A 95:4635-4640.

Scheid MP, Woodgett JR (2001) PKB/AKT: functional insights from genetic models. Nat Rev Mol Cell Biol 2:760-768.

Schindler C (1999) Cytokines and JAK-STAT signaling. Exp Cell Res 253:7-14.

Schindler C, Strehlow I (2000) Cytokines and STAT signaling. Adv Pharmacol 47:113-174.

Schubert M, Gautam D, Surjo D, Ueki K, Baudler S, Schubert D, Kondo T, Alber J, Galldiks N, Kustermann E, Arndt S, Jacobs AH, Krone W, Kahn CR, Bruning JC (2004) Role for neuronal insulin resistance in neurodegenerative diseases. Proc Natl Acad Sci U S A 101:3100-3105.

Schwaiger FW, Hager G, Schmitt AB, Horvat A, Streif R, Spitzer C, Gamal S, Breuer S, Brook GA, Nacimiento W, Kreutzberg GW (2000) Peripheral but not central axotomy induces changes in Janus kinases (JAK) and signal transducers and activators of transcription (STAT). Eur J Neurosci 12:1165-1176.

Sinor AD, Greenberg DA (2000) Erythropoietin protects cultured cortical neurons, but not astroglia, from hypoxia and AMPA toxicity. Neurosci Lett 290:213-215.

Siren AL, Fratelli M, Brines M, Goemans C, Casagrande S, Lewczuk P, Keenan S, Gleiter C, Pasquali C, Capobianco A, Mennini T, Heumann R, Cerami A, Ehrenreich H, Ghezzi P (2001) Erythropoietin prevents neuronal apoptosis after cerebral ischemia and metabolic stress. Proc Natl Acad Sci U S A 98:4044-4049.

Socolovsky M, Fallon AE, Wang S, Brugnara C, Lodish HF (1999) Fetal anemia and apoptosis of red cell progenitors in Stat5a-/-5b-/- mice: a direct role for Stat5 in Bcl-X(L) induction. Cell 98:181-191.

Solaroglu I, Solaroglu A, Kaptanoglu E, Dede S, Haberal A, Beskonakli E, Kilinc K (2003) Erythropoietin prevents ischemia-reperfusion from inducing oxidative damage in fetal rat brain. Childs Nerv Syst 19:19-22.

Stoica BA, Movsesyan VA, Lea PMt, Faden AI (2003) Ceramide-induced neuronal apoptosis is associated with dephosphorylation of Akt, BAD, FKHR, GSK-3beta, and induction of the mitochondrial-dependent intrinsic caspase pathway. Mol Cell Neurosci 22:365-382.

Sui X, Krantz SB, You M, Zhao Z (1998) Synergistic activation of MAP kinase (ERK1/2) by erythropoietin and stem cell factor is essential for expanded erythropoiesis. Blood 92:1142-1149.

Trapp T, Korhonen L, Besselmann M, Martinez R, Mercer EA, Lindholm D (2003) Transgenic mice overexpressing XIAP in neurons show better outcome after transient cerebral ischemia. Mol Cell Neurosci 23:302-313.

Uddin S, Kottegoda S, Stigger D, Platanias LC, Wickrema A (2000) Activation of the Akt/FKHRL1 pathway mediates the antiapoptotic effects of erythropoietin in primary human erythroid progenitors. Biochem Biophys Res Commun 275:16-19.

Verdier F, Chretien S, Muller O, Varlet P, Yoshimura A, Gisselbrecht S, Lacombe C, Mayeux P (1998) Proteasomes regulate erythropoietin receptor and signal transducer and activator of transcription 5 (STAT5) activation. Possible involvement of the ubiquitinated Cis protein. J Biol Chem 273:28185-28190.

Villa P, Bigini P, Mennini T, Agnello D, Laragione T, Cagnotto A, Viviani B, Marinovich M, Cerami A, Coleman TR, Brines M, Ghezzi P (2003) Erythropoietin selectively attenuates cytokine production and inflammation in cerebral ischemia by targeting neuronal apoptosis. J Exp Med 198:971-975.

Weishaupt JH, Rohde G, Polking E, Siren AL, Ehrenreich H, Bahr M (2004) Effect of erythropoietin axotomy-induced apoptosis in rat retinal ganglion cells. Invest Ophthalmol Vis Sci 45:1514-1522.

Yasuda Y, Nagao, M., Okano, M., Masuda, S., Sasaki, R., Konishi, H., Tanimura, T. (1993) Localization of erythropoietin and erythropoietin-receptor in postimplantation mouse embryos. Develeop Growth&Differ 36:711-722.

Yousouffian H, Longmore, G., Neumann, D., Yoshimura, A., Lodish, HF (1993) Structure, Function, and activation of the erythropoietin receptor. Blood 81:2223-2236.

Yu X, Shacka JJ, Eells JB, Suarez-Quian C, Przygodzki RM, Beleslin-Cokic B, Lin CS, Nikodem VM, Hempstead B, Flanders KC, Costantini F, Noguchi CT (2002) Erythropoietin receptor signalling is required for normal brain development. Development 129:505-516.

Zhou H, Li XM, Meinkoth J, Pittman RN (2000) Akt regulates cell survival and apoptosis at a postmitochondrial level. J Cell Biol 151:483-494.

Chapter 4

REGULATION OF ERYTHROPOIETIN EXPRESSION IN THE NERVOUS SYSTEM: THE HYPOXIA INDUCIBLE FACTOR

Juan C. Chavez and JoAnn M. Gensert
Burke/Cornell Medical Research Institute, Department of Neurology and Neuroscience, Weill Medical College of Cornell University, 785 Mamaroneck Avenue, White Plains, NY 10605, USA.

Abstract: Hypoxia is associated with a variety of CNS diseases including stroke, traumatic brain injury and spinal cord injury. Induction of erythropoietin (EPO) is a physiological response to hypoxia in oxygen-deprived tissues. Hypoxia-induced epo gene expression is regulated by the transcriptional activator hypoxia inducible factor (HIF). The epo gene contains a HIF binding site in its 3' untranslated region, and its expression is upregulated concomitantly with HIF activation in a variety of cell culture models as well as in the brains of rodents exposed to hypoxia or pharmacological agents that mimic hypoxia, such as iron chelators and cobalt chloride. Since HIF is a master regulator of oxygen homeostasis in all mammalian cells and controls the expression of a variety of genes required for cellular adaptation to hypoxia (including epo), this review covers the current knowledge about the oxygen sensing mechanism that regulates the activation of HIF under hypoxic conditions. An important challenge for the future is to determine how modulating the activation of HIF with the subsequent expression of EPO can be beneficial for neural survival under stress conditions that involve hypoxia.

Key words: hypoxia, transcription, ischemia, oxygen, neuroprotection.

1. INTRODUCTION

The mammalian central nervous system (CNS) requires a steady supply of oxygen (O_2) to support synthesis of ATP, which is essential to maintain normal cell homeostasis, in particular neuronal function (Siesjo and Plum, 1971; Siesjo, 1984). When the availability of O_2 becomes limited, the CNS is subject to a detrimental metabolic stress that may cause cell death depending on the duration and severity of the insult (Katsura et al., 1994; Siesjo, 1981). This hypoxic stress causes the activation of a variety of endogenous compensatory mechanisms aimed at restoring the balance between local oxygen delivery and tissue oxygen consumption. Immediate brain adaptive responses to hypoxia include vasodilation of cerebral arteries and veins resulting in a reversible increase in blood flow (LaManna et al., 1992; LaManna and Harik, 1997); whereas long-term responses involve changes at the level of gene expression. The hypoxia inducible factors (HIFs) are central regulators of these long-term adaptive responses (Semenza, 1998; Semenza, 1999).

Hypoxia inducible factors control the hypoxia-dependent upregulation of a variety of target genes that together facilitate the cellular adaptation to low O_2 (Semenza, 2000a). These hypoxia-inducible genes include glucose transporters (Glut-1) and glycolytic enzymes, which will promote metabolic adaptation (Semenza et al., 1994), and angiogenic factors such as the vascular endothelial growth factor (VEGF), which will stimulate formation of new blood vessels and augment tissue O_2 delivery (Forsythe et al., 1996; Semenza, 2003a). In addition, at the systemic level, hypoxia induces the expression of EPO that will enhance red blood cell production and augment O_2 carrying capacity (Monge and Leon-Velarde, 1991; Semenza, 1994b). Hypoxia also induces expression of EPO in the CNS where it functions as a trophic factor promoting cell survival, particularly during severe insults such as stroke and traumatic brain injury (Buemi et al., 2002; Digicaylioglu and Lipton, 2001; Gassmann et al., 2003; Marti, 1996; Marti et al., 1997).

Erythropoietin is the best characterized oxygen-responsive gene and an ideal model to study hypoxia induced gene expression at the molecular level (Bunn et al., 1998). In fact, our current knowledge on the mechanism of oxygen sensing that controls HIF activation and oxygen-regulated gene expression in mammalian cells was acquired primarily through our efforts to understand how EPO expression is increased during hypoxia (Goldberg et al., 1991; Semenza, 1994b; Semenza, 1994a; Wang and Semenza, 1995; Wang and Semenza, 1996). Hypoxia-induced expression of EPO is controlled primarily at the level of transcription; however, mRNA stabilization contributes significantly to the total amount of EPO message (Ho et al., 1995). Within the epo gene, there is a cis-acting regulatory element that is responsible for its hypoxic induction (Maxwell et al., 1993). Characterization of this 3' enhancer led to the discovery of HIF, which binds to a specific region within this enhancer thereby mediating transcriptional activation of EPO (Maxwell et al., 1993; Semenza and Wang, 1992; Wang and Semenza, 1995). This discovery was central to the understanding of the molecular mechanisms regulating hypoxia-induced gene expression. It unveiled a ubiquitous oxygen sensing mechanism that regulates the activation of HIF, which not only regulates the expression of EPO but also a broad range of genes that together facilitate the cellular adaptation to low oxygen. This mechanism is highly conserved in vertebrates and can be found also in lower organisms such as *Drosophila Melanoghaster* and *Caenorhabditis Elegans* (Bacon et al., 1998; Epstein et al., 2001; Semenza, 2001). This chapter will discuss the role of HIF as a key transcriptional regulator of EPO and the oxygen sensing mechanism that activates HIF in mammalian tissues including the CNS.

2. ERYTHROPOIETIN EXPRESSION IN THE CNS

Systemic EPO is produced mainly in the liver during fetal development and then in the kidney in adulthood (Zanjani et al., 1981). The expression of EPO, however, is not restricted to these organs. EPO also is expressed in testis, uterus, lung, spleen, heart and bone marrow when animals are subjected to hypoxia (Fandrey and Bunn, 1993; Tan et al., 1991). Moreover, EPO and EPO receptor (EPOR) are expressed in the developing and mature central nervous system (CNS) (Digicaylioglu et al., 1995; Fandrey and Bunn, 1993; Marti et al., 1996). In early fetal stages, EPO expression has been detected in the periventricular germinal matrix zone, subpial granular layer, thalamus, hippocampus, lateral geniculate nuclei, cortex and spinal cord (Dame et al., 2000; Juul et al., 1998). In the adult brain, constitutive expression of EPO mRNA has been detected in cortical regions, amygdala and hippocampus – areas particularly susceptible to hypoxic/ischemic insults (Marti et al., 1996; Marti et al., 1997). At the cellular level, EPO and/or EPOR have been detected in all major neural cell types. EPO and EPOR expression *in vitro* and *in vivo* in astrocytes, with the receptor robustly expressed in astrocyte processes surrounding capillaries and in neurons is well established (Bernaudin et al., 2000; Marti et al., 1996; Masuda et al., 1994; Siren et al., 2001). Oligodendrocyte expression of EPO and EPOR has been reported in cells isolated from embryonic rat brain and human CNS (Nagai et al., 2001; Sugawa et al., 2002). Although expression of EPOR has been reported for microglial cells isolated from human tissue (Nagai et al., 2001), as well as for human and rodent endothelial cells *in vitro* and *in vivo* (Anagnostou et al., 1994; Brines et al., 2000; Yamaji et al., 1996), whether these cells produce EPO remains to be established. As in liver and kidney, EPO expression in the CNS is also oxygen-sensitive. Albeit, the temporal pattern of hypoxia-induced EPO expression in CNS differs from that in kidney and liver; whereas EPO expression in response to hypoxia is transient in these tissues, EPO expression in the brain is sustained for as long as the hypoxic stimulus persists (Chikuma et al., 2000). Hypoxia induces EPO in both astrocytes and neurons both at the mRNA and protein levels (Bernaudin et al., 2000; Masuda et al., 1994). Astrocytic and neuronal EPO expression is induced also by desferrioxamine (DFO) and cobalt chloride ($CoCl_2$), two classic pharmacologic agents that mimic hypoxia and cause activation of HIF (Bergeron et al., 2000).

3. THE EPO GENE: PROMOTER REGION AND CIS-ACTING REGULATORY ELEMENTS.

Epo is a single-copy gene located on chromosome 7 in humans and on chromosome 5 in mice (Bunn and Poyton, 1996; Bunn et al., 1998). This gene consists of five exons, four introns and several cis-acting regulatory elements (Fig. 4-1). These DNA regulatory sequences were identified through a transgenic approach in which various DNA fragments of the human epo gene were introduced, with subsequent EPO induction examined in different tissues (Semenza, 1994a). These studies identified specific hypoxic responsive regions of the epo gene for the liver and kidney as well as negative regulatory elements that repressed its expression in a tissue specific manner (Semenza, 1994b; Semenza, 1994a). For further analysis of hypoxia-induced EPO gene expression, two human liver tumor cell lines, HepG2 and Hep3B, were utilized. These cell lines express high levels of EPO when challenged with hypoxia and serve as classical cellular model systems to study the molecular mechanisms of hypoxia-induced EPO expression (Goldberg et al., 1991; Nielsen et al., 1987). Two neuroblastoma cell lines, SH-SY5Y and Kelly, can also be used as models to study hypoxia-induced EPO expression in neural cells. These cell lines produce EPO in response to hypoxia, although they require a more severe hypoxic stress (lower pO_2) compared to hepatoma cells (Stolze et al., 2002).

The epo promoter, unlike most promoters, lacks the canonical TATA or CAAT sequences. Although this promoter is weak, during hypoxia, its activity increases, synergizing with other cis-acting elements to achieve a robust induction of epo gene expression *in vivo* (several hundred- to 1000-fold) and reporter gene expression *in vitro* (50- to 100-fold) (Blanchard et al., 1992; Blanchard et al., 1993; Imagawa et al., 1991). Hypoxia inducible tissue specific cis-regulatory elements for liver (liver inducible element, LIE) and for kidney (kidney inducible element, KIE) are located respectively in the 3'- and 5'- flanking regions of the epo gene. A negatively regulated liver element (NRLE) is located 3' to the LIE, and a negative regulatory element (NRE) that represses epo gene expression in non-EPO expressing cells is located just upstream of the transcription start codon (Semenza et al.,

1990; Semenza et al., 1991a). Identification of neural-specific cis-elements has not been elucidated.

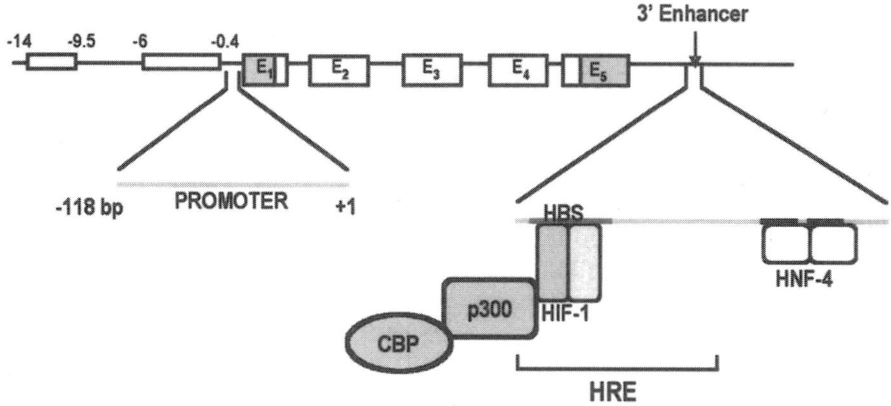

Figure 4-1. The human epo gene including the promoter and 3' enhancer region (Adapted from Bunn et. al., 1999)

A region known as the 3' enhancer is the most critical hypoxia-responsive cis-element of the epo gene. It is a conserved 40-bp element in the 3'-region, located 120 bp downstream of the polyadenylation site in a region highly conserved between human and mouse sequences (Beck et al., 1991; Pugh et al., 1991; Pugh et al., 1994a; Semenza et al., 1991b; Semenza and Wang, 1992) . This cis-acting DNA element, when linked to a reporter gene and transfected into hepatoma cells, is sufficient to elicit a hypoxia-induced response similar to the EPO response *in vivo* (Maxwell et al., 1993). There are three distinct regions of this enhancer: a HIF-1 binding site (HBS), a CACA repeat, and a direct repeat of two steroid receptor half sites (DR-2) (Blanchard et al., 1992; Pugh et al., 1994b; Semenza and Wang, 1992). The HBS consists of a highly conserved 5' portion element (13 bp) with a consensus sequence CA/(G)CGTGCT. Electromobility shift assays using double stranded oligonucleotides were used to demonstrate that a nuclear protein (HIF-1) from hepatoma cells binds this sequence only when cells are subjected to hypoxia; this binding is essential for hypoxic EPO induction (Semenza and Wang, 1992). In addition, a constitutive protein complex formed by ATF-1 and CREB-1 was identified (Kvietikova et al., 1995). However, the role of this complex in hypoxia-regulated EPO expression has not yet been elucidated. A second element of the 3' enhancer in the human gene consists of three CA repeats (Pugh et al., 1991; Pugh et al., 1994b). Although not highly conserved, it is necessary for hypoxic induction of EPO as well as of other HIF-1 target genes. Which protein(s) recognize and bind this CA repeat site is not yet known. The third element of the epo enhancer, DR-2, is located 3' to the other two and is absolutely required for the induction of EPO expression during hypoxia (Blanchard et al., 1992). Direct repeat steroid receptor half sites are known to bind a variety of hormone nuclear receptors. However, none of the classic ligands for the direct repeats half sites was shown to modulate epo expression through the DR-2 binding sites (Blanchard et al., 1992). Screening for orphan receptors has identified a protein called hepatocyte nuclear factor-4α (HNF4α). In the Hep3B hepatoma cell line, HNF4α is required for the induction of EPO during hypoxia (Galson et al., 1994). The neuroblastoma cell lines, SH-SY5Y and Kelly, which show robust upregulation of EPO in response to hypoxia, however, lack HNF4α and other HNF4 isoforms (Stolze et al., 2002). Therefore the DR-2 sites and HNF4α may play a role in tissue specific expression of EPO. Since the DR-2 sites are required for hypoxic induction of EPO, a yet unknown factor must substitute HNF-4 and regulate EPO expression in neural cells.

As noted previously, astrocytes and neurons upregulate EPO *in vitro* and *in vivo* not only in response to hypoxia, but also in response to DFO and $CoCl_2$. As these two agents are well-characterized HIF-1 activators, it is likely that HIF-1 is a central mediator of hypoxia-induced EPO expression in neural cells. Although details about other regulatory elements that contribute to hypoxia-induced EPO expression in the CNS are not known, the oxygen sensing mechanism that regulates HIF-1 activation is well characterized and it is highly conserved in all tissues including the CNS.

4. HYPOXIA INDUCIBLE FACTOR (HIF)

HIF is a heterodimer formed by two subunits that belong to the PAS (Per, Arnt, Sim) family of basic helix-loop-helix (bHLH) transcription factors; these subunits are designated HIF-α and HIF-β. The expression of the HIF-α subunits is regulated by oxygen levels, whereas the HIF-β subunits, also known as arylhydrocarbon receptor nuclear translocator (ARNT), are constitutive nuclear proteins that dimerize with other bHLH-PAS transcription factors (Semenza et al., 1997; Semenza, 1998; Wang and Semenza, 1995; Wang et al., 1995). Currently, three HIF-α subunits (HIF-1α, HIF-2α/EPAS1 and HIF-3α) as well as three HIF-β subunits (HIF-1β/ARNT1, ARNT2 and ARNT3) are known (Semenza, 1999; Talks et al., 2000). The most widely expressed alpha subunit in mammalian tissues is HIF-1α; indeed most of our knowledge about HIF comes from studies of the mechanism regulating the expression of HIF-1α protein and the transcriptional activity of the HIF-1 complex (HIF-1α/HIF-1β heterodimer).

The other HIF-α subunits are regulated by a similar mechanism although they appear to have more specialized and tissue specific functions (Semenza, 1999). During normoxia, the HIF-1α protein is constitutively expressed, but it is rapidly destroyed by the ubiquitin-proteosome system, such that almost no HIF-1α protein accumulates (Huang et al., 1998; Salceda and Caro, 1997; Salceda and Caro, 1997). Under hypoxic conditions, degradation of the HIF-1α subunit is prevented, allowing HIF-1α to accumulate within the nucleus where it dimerizes with HIF-1β forming the HIF-1 transcriptional complex (Jewell et al., 2001; Salceda and Caro, 1997). HIF-1 binds to a consensus DNA sequence A/(G)CGTG within the hypoxia response elements (HRE) of numerous hypoxic responsive target genes that include EPO, glycolytic enzymes, angiogenic factors, and glucose transporters, among others (Semenza, 1999; Wang and Semenza, 1993; Wang et al., 1995).

4.1 Molecular mechanism of HIF-1 activation during hypoxia: oxygen-sensing mechanism.

Regulation of HIF by O_2 is mediated by two distinct pathways that involve enzymatic trans-4-hydroxylation of two proline residues and the β-hydroxylation of an asparagine residue in the HIF-α subunits. Prolyl hydroxylation the HIF-α subunit is carried out by an enzyme encoded by the egg-laying abnormal-9 (*Egl-9*) gene in *C. Elegans* and *D. Melanogaster*. The mammalian homologues are named egg laying nine 1 (EGLN1), EGLN2 and EGLN3, also called prolyl hydroxylase domain-containing proteins PHD2, PHD1, and PHD3, respectively (Epstein et al., 2001). Degradation of HIF-α under normoxic conditions is triggered by post-translational hydroxylation of the conserved proline residues, Pro-402 and Pro-564, within a region of the HIF-α protein known as the oxygen-dependent degradation (ODD) domain (see below and Fig. 4-3). The hydroxylated proline residues in this domain are recognized by the product of the von Hippel-Lindau tumor suppressor gene (pVHL), which acts as the recognition component of a multiprotein ubiquitin E3 ligase complex, thus targeting HIF-α to ubiquitin-mediated proteolysis in the proteosome (Ivan et al., 2001; Jaakkola et al., 2001; Masson et al., 2001; Maxwell et al., 1999). This regulatory post-translational modification is inherently oxygen dependent, since the hydroxyl group is derived from molecular oxygen (Bruick and McKnight, 2001; Epstein et al., 2001; Ivan et al., 2001). The prolyl hydroxylation also requires the cofactors 2-oxoglutarate, vitamin C and iron (II). The requirement of iron explains the hypoxic-mimetic effects of iron chelators (such as deferoxamine mesylate) and iron antagonists (such as cobalt chloride). Under low O_2 conditions, or in the presence of iron chelators, HIF-α is not hydroxylated by PHDs and therefore HIF-α is neither recognized by pVHL nor targeted for degradation by the proteosome (Fig. 4-2). As a result, HIF-α accumulates in the nucleus and is available to dimerize with HIF-β subunits to form the active HIF complex that activates transcription of target genes including EPO.

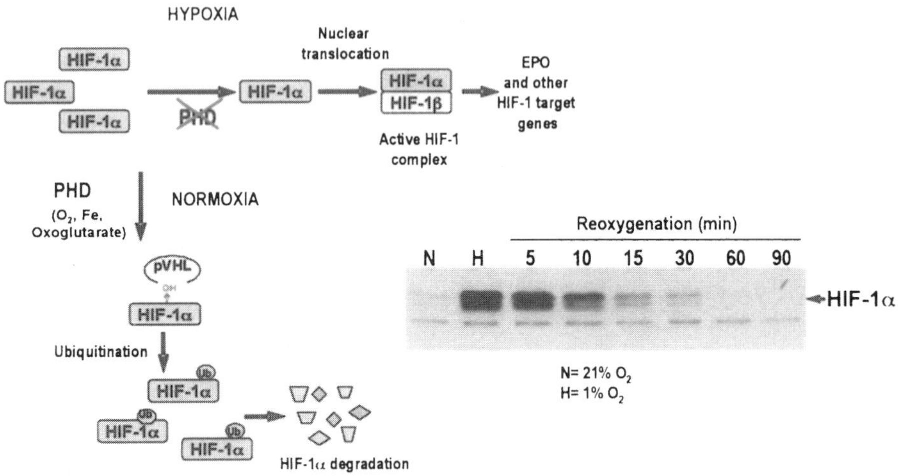

Figure 4-2. Regulation of HIF-1α by O_2 level. In cultured cortical neurons, HIF-1α protein accumulation is regulated also by O_2 tension as shown by western blot analysis.

Hypoxia affects not only HIF-α protein stability but also the transcriptional activity of the HIF complex (Bruick and McKnight, 2002; Jiang et al., 1997; Pugh et al., 1997). The three alpha subunits contain two transactivation domains (TAD) that interact with transcriptional co-activators essential for gene expression (Figure 3). The amino terminal TAD (N-TAD, aa 531-575) overlaps with the ODD domain (aa 401-603). The carboxy terminal TAD (C-TAD, aa 786-826) is independent of the ODD domain and is able to recruit co-activators such as p300/CBP under hypoxic conditions only (Ema et al., 1999; Gu et al., 2001; Semenza, 2002). The regulation of this C-TAD involves an oxygen-dependent hydroxylation of a conserved asparagine residue (Asn-803 in HIF-1α and Asn-851 in HIF-2α). This hydroxylation is catalyzed by a novel asparaginyl hydroxylase that was described previously as a factor inhibiting HIF-1 (FIH-1). This asparaginyl hydroxylase is also a member of the 2-oxoglutarate- and iron-dependent dioxygenase superfamily; hence it is inhibited by hypoxia (Lando et al., 2002;Mahon et al., 2001). When the asparagine residue is hydroxylated, the C-TAD cannot interact with the transcriptional coactivator p300/CBP and the HIF-1 transcriptional activity is reduced.

Histone acetylation state is also a factor that affects HIF-1 transcriptional activity. In addition to promoting ubiquitination and degradation of HIF-1α, VHL forms a ternary complex with HIF-1α and the co-repressor FIH-1. Both VHL and FIH-1 recruit histone deacetylases that may contribute to the loss of HIF-1 transcriptional activity under non-hypoxic conditions (Mahon et al., 2001).

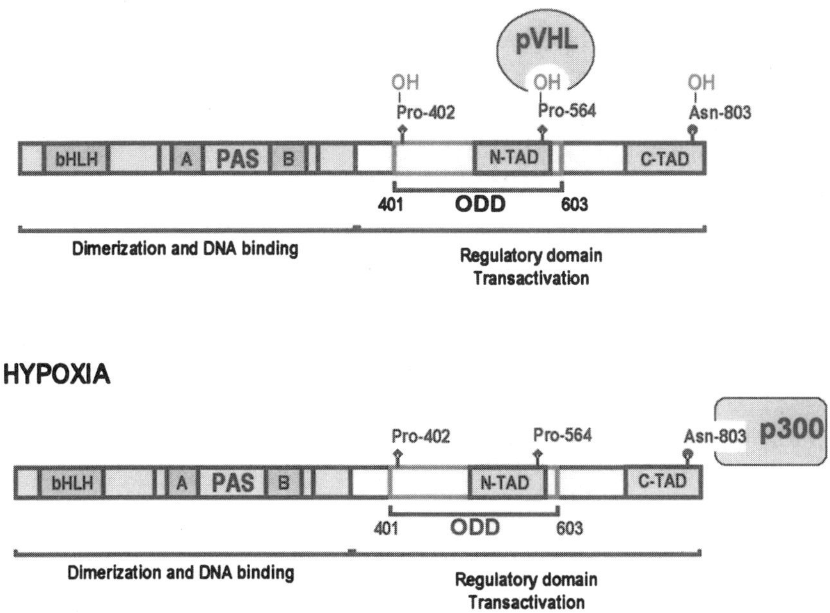

Figure 4-3. Oxygen mediated post-translational modifications of HIF-1α subunit.

In addition to hydroxylation, a new post-translational modification of the HIF-1α subunit has been described, where a novel HIF-1α protein acetyl transferase called ARD1 was shown to acetylate the lysine 532 residue of HIF-1α. This acetylation enhances the interaction of HIF-1α with VHL, thereby augmenting the subsequent HIF-1α ubiquitination and proteosomal degradation. Moreover, this report showed that the expression of ARD1 is reduced by hypoxia, consistent with the accumulation of HIF-1α during low oxygen conditions (Jeong et al., 2002). Collectively, these findings demonstrate that during hypoxia, HIF-1 is regulated by a carefully controlled signal transduction pathway in which a group of hydroxylases act as putative oxygen sensors by way of their requirement of oxygen for activity. These hydroxylases catalyze unique oxygen dependent posttranslational modifications of the HIF-1α subunit that control its degradation and regulate HIF transcriptional activity.

4.2 Regulation of prolyl hydroxylases expression

Little is known about the expression and functions of HIF-1 prolyl hydroxylases in the CNS. However, the regulation of HIF by the PHDs known to date, PHD1, PHD2, and PHD3, has been studied *in vitro* (in chemical assays) and in the context of a variety of non-neural cell types. All three of the PHD isoforms hydroxylate HIF-α peptides *in vitro* (Bruick and McKnight, 2001; Epstein et al., 2001) and contribute to the regulation of HIF in a variety of cell contexts (Appelhoff et al., 2004; Hirsila et al., 2003). Furthermore, when overexpressed in cells, all three PHDs can suppress HRE-mediated reporter gene activity (Huang et al., 2002; Metzen et al., 2003). Despite distinct intracellular localization patterns of exogenous PHDs – PHD1 is exclusively nuclear, PHD2, mainly cytoplasmic, and PHD3, both cytoplasmic and nuclear – each of the PHDs can regulate nuclear HIF-1α during hypoxia (Metzen et al., 2003). Metzen and colleagues also showed that endogenous PHD2 mRNA and PHD3 mRNA are hypoxia-induced. However, during normoxia, a dominant role for endogenous PHD2 has been demonstrated, in a variety of non-neural cell lines (Berra et al., 2003). Furthermore, PHD2 has a greater influence on HIF-1α than on HIF-2α; whereas for PHD3, the opposite pattern emerges (Appelhoff et al., 2004). Admittedly, variations in expression levels of endogenous PHD isoforms appear to be cell type- and culture condition- dependent, warranting investigation of PHDs in neural cell types. To further understand HIF regulation of EPO in the brain, study of these critical HIF regulators, the PHDs, in a neural cell context is essential. These observations suggest that a novel feedback mechanism for adjusting hypoxia-induced gene expression exist that involves regulation of PHD expression.

In a recent study, Nakayama et al. had identified a novel mechanism that regulates the availability of PHD1 and PHD3 and consequently affects the abundance of HIF-1α (Nakayama et al., 2004). This study showed that PHD1 and PHD3 protein levels are regulated by members of the E3 ubiquitin ligase family Siah2 and Siah1a. These E3 ligases mediate the ubiquitination and the proteosome-dependent degradation of PHD1/3 during hypoxia. Generally, PHD activity is diminished during hypoxia, although even at low oxygen concentrations, residual PHD activity may persist. To overcome this residual activity, an additional mechanism of PHD regulation at the protein level is required during hypoxia to facilitate HIF-1 activation and upregulation of hypoxia-responsive genes including EPO (Simon, 2004). Regulation of PHD availability is therefore another step in the complex pathway that regulates HIF activation.

4.3 Alternate mechanisms stabilize HIF-1α under normoxic conditions

As noted above, DFO and $CoCl_2$ are classic pharmacological agents that mimic some cellular hypoxic responses, including activation of HIF-1. Numerous studies have shown that a variety of iron chelators including mimosine and DFO are potent activators of HIF since PHD activity requires iron (Ivan et al., 2002; Warnecke et al., 2003). In addition, the divalent transition metal Cobalt (Co^{+2}) has been shown to induce HIF-1α protein accumulation under normoxic conditions. Although it was initially proposed that Co^{+2} acts by displacing iron, a recent report demonstrated that Co^{+2} induces HIF-1α protein accumulation by disrupting the interaction between hydroxylated HIF-1α and VHL, consequently preventing HIF-1α ubiquitination and proteosomal degradation (Yuan et al., 2003). Both DFO and $CoCl_2$ have been used successfully to activate HIF and the expression of HIF target genes in neonatal rat brain and adult mouse and rat brain (Bergeron et al., 2000) and were effective in reducing brain injury associated with ischemia in different animal models (Sharp and Bernaudin, 2004).

HIF-1α protein expression, HIF-1 DNA binding activity and HIF-1 target gene expression under non-hypoxic conditions are induced also by a variety of growth factors and cytokines, including epidermal growth factor (EGF), fibroblast growth factor 2 (FGF-2), insulin, insulin-like growth factor 1 and 2 (IGF-1, IGF-2), tumor necrosis factor-α (TNFα), interleukin-1β and angiotensin II (Semenza et al., 2000; Semenza, 2000c; Semenza, 2000b). These growth factors and cytokines bind their cognate receptor tyrosine kinases and activate a variety of signaling pathways, including the phosphatidylinositol 3-kinase (PI3K), the serine-threonine protein kinase Akt (protein kinase B), the mammalian target of rapamycin (mTOR, also known as FRAP) and the ERK/MAPK pathway (Semenza, 2000c; Semenza, 2003b). Many of these pathways have been implicated in the growth factor mediated activation of HIF-1 in a variety of cell lines.

In the CNS, IGF-1 is so far the only growth factor that has been shown to activate the HIF pathway. Interestingly, IGF-1, IGF-2 and insulin can stimulate EPO production in primary cultured astrocytes (Masuda et al., 1997). In a model of global cerebral ischemia in rats, IGF-1 mediates in part the activation of HIF-1 independently of hypoxia. Moreover, exogenous systemic or intra-cerebroventricular infusion of IGF causes HIF-1α accumulation and expression of HIF-1 target genes including EPO (Chavez and LaManna, 2002). In a recent study, Lopez-Lopez et al. showed that IGF-1 induces the growth of cultured brain endothelial cells through activation of HIF and its target gene, VEGF. This study also showed that systemic injection of IGF-I in adult mice increases brain vessel density (Lopez-Lopez et al., 2004). Taken together these data support the role of IGF-1 as an important regulator of HIF activation in the adult CNS.

4.4 Physiologic role of HIF-1 in the CNS

Hypoxia inducible factor-1 has a critical physiological role in the CNS; it is absolutely required for normal development. Mouse embryos that lack HIF-1α die at midgestation, with multiple cardiovascular defects and mesenchymal cell death (Yu et al., 1999) . Also, HIF-1α is necessary for normal development of the brain. In a mouse model of neural cell specific HIF-1α-deficiency, animals were viable and reached adulthood; however, they developed hydrocephalus and showed a marked reduction in brain mass (Tomita et al., 2003). In the adult brain, HIF-1α is expressed constitutively (Stroka et al., 2001) and is further induced by hypoxia in neurons, astrocytes, ependymal cells and possibly endothelial cells (Chavez et al., 2000b). Whether HIF-1 is activated in microglia and oligodendrocytes is not known. In contrast, HIF-2α seems to be induced preferentially in glia and endothelial cells, but not in neurons (Wiesener et al., 2003). A recent study suggests that expression of HIF-1α and HIF-2α results in the induction of different HIF target genes. In particular, the expression of EPO seems to depend primarily on HIF-2α activation (Ralph et al., 2004). In the brains of rodents exposed to hypoxia, HIF-1α accumulation correlates with the upregulation of HIF regulated genes that include glucose transporters (Glut-1), glycolytic enzymes, pro-angiogenic factors (VEGF and Flt-1) and EPO (Bergeron et al., 1999;Chavez et al., 2000a) (Fig. 4-4). Similarly, in a variety of cerebral ischemia models, EPO is upregulated at the mRNA level (Bergeron et al., 1999;Bergeron et al., 2000;Sharp et al., 2001;Sharp and Bernaudin, 2004). Taken together, the *in vivo* and *in vitro* data implicate HIF as a critical mediator of EPO expression in the CNS.

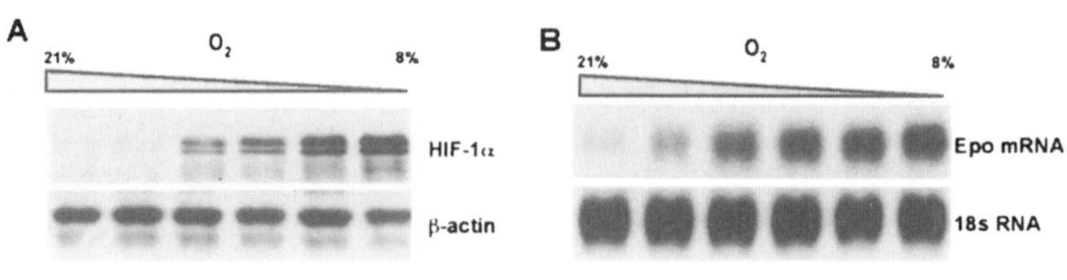

Figure 4-4. Oxygen-mediated accumulation of HIF-1α protein and EPO mRNA upregulation in mouse cerebral cortex

5. CONCLUDING REMARKS

This chapter emphasizes the oxygen sensing mechanism that regulates HIF activation and the expression of its target genes, including EPO. HIF is considered the master regulator of O_2 homeostasis in all mammalian cells including neural cells, and therefore it is likely that HIF is a critical regulator of hypoxia-induced EPO expression in the CNS. Indeed, the epo gene contains a HIF binding site, and is upregulated concomitantly with HIF activation in the brain of rodents exposed to hypoxia (Fig. 4-4) or pharmacological agents that mimic hypoxia, such as iron chelators and cobalt chloride (Bergeron et al., 2000). Besides HIF, probably other transcriptional regulators, activators and/or repressors, and perhaps other yet unknown neural specific cis-acting DNA elements, participate in the control of CNS specific EPO expression.

Since hypoxia is associated with a variety of CNS diseases, including stroke, traumatic brain injury and spinal cord injury, it is likely that HIF activation and EPO upregulation also occurs in these pathologies. In fact, upregulation of EPO expression in ischemic brain tissue has been documented in a variety of animal models of cerebral ischemia (Sharp and Bernaudin, 2004).

Despite the lack of mechanistic studies regarding the role of HIF in CNS disease, many laboratories have shown that different treatments that target HIF activation are beneficial in a variety of CNS injury models including cerebral ischemia (Sharp and Bernaudin, 2004). Furthermore, there is compelling evidence supporting the role of EPO as a potent survival factor in various models of CNS injury *in vitro* and *in vivo* (Digicaylioglu and Lipton 2001). An important challenge for the future is to determine whether HIF or EPO can be effectively manipulated pharmacologically to promote neural survival under stress conditions associated with hypoxia.

REFERENCES

Anagnostou A, Liu Z, Steiner M, Chin K, Lee ES, Kessimian N, Noguchi CT (1994) Erythropoietin receptor mRNA expression in human endothelial cells. Proc Natl Acad Sci U S A 91: 3974-3978.

Appelhoff RJ, Tian YM, Raval RR, Turley H, Harris AL, Pugh CW, Ratcliffe PJ, Gleadle JM (2004) Differential function of the prolyl hydroxylases PHD1, PHD2, and PHD3 in the regulation of hypoxia-inducible factor. J Biol Chem 279: 38458-38465.

Bacon NC, Wappner P, O'Rourke JF, Bartlett SM, Shilo B, Pugh CW, Ratcliffe PJ (1998) Regulation of the Drosophila bHLH-PAS protein Sima by hypoxia: functional evidence for homology with mammalian HIF-1 alpha. Biochem Biophys Res Commun 249: 811-816.

Beck I, Ramirez S, Weinmann R, Caro J (1991) Enhancer element at the 3'-flanking region controls transcriptional response to hypoxia in the human erythropoietin gene. J Biol Chem 266: 15563-15566.

Bergeron M, Gidday JM, Yu AY, Semenza GL, Ferriero DM, Sharp FR (2000) Role of hypoxia-inducible factor-1 in hypoxia-induced ischemic tolerance in neonatal rat brain. Ann Neurol 48: 285-296.

Bergeron M, Yu AY, Solway KE, Semenza GL, Sharp FR (1999) Induction of hypoxia-inducible factor-1 (HIF-1) and its target genes following focal ischaemia in rat brain. Eur J Neurosci 11: 4159-4170.

Bernaudin M, Bellail A, Marti HH, Yvon A, Vivien D, Duchatelle I, MacKenzie ET, Petit E (2000) Neurons and astrocytes express EPO mRNA: oxygen-sensing mechanisms that involve the redox-state of the brain. Glia 30: 271-278.

Berra E, Benizri E, Ginouves A, Volmat V, Roux D, Pouyssegur J (2003) HIF prolyl-hydroxylase 2 is the key oxygen sensor setting low steady-state levels of HIF-1alpha in normoxia. EMBO J 22: 4082-4090.

Blanchard KL, Acquaviva AM, Galson DL, Bunn HF (1992) Hypoxic induction of the human erythropoietin gene: cooperation between the promoter and enhancer, each of which contains steroid receptor response elements. Mol Cell Biol 12: 5373-5385.

Blanchard KL, Fandrey J, Goldberg MA, Bunn HF (1993) Regulation of the erythropoietin gene. Stem Cells 11 Suppl 1: 1-7.

Brines ML, Ghezzi P, Keenan S, Agnello D, de Lanerolle NC, Cerami C, Itri LM, Cerami A (2000) Erythropoietin crosses the blood-brain barrier to protect against experimental brain injury. Proc Natl Acad Sci U S A 97: 10526-10531.

Bruick RK, McKnight SL (2001) A conserved family of prolyl-4-hydroxylases that modify HIF. Science 294: 1337-1340.

Bruick RK, McKnight SL (2002) Transcription. Oxygen sensing gets a second wind. Science 295: 807-808.

Buemi M, Cavallaro E, Floccari F, Sturiale A, Aloisi C, Trimarchi M, Grasso G, Corica F, Frisina N (2002) Erythropoietin and the brain: from neurodevelopment to neuroprotection. Clin Sci (Lond) 103: 275-282.

Bunn HF, Gu J, Huang LE, Park JW, Zhu H (1998) Erythropoietin: a model system for studying oxygen-dependent gene regulation. J Exp Biol 201 (Pt 8): 1197-1201.

Bunn HF, Poyton RO (1996) Oxygen sensing and molecular adaptation to hypoxia. Physiol Rev 76: 839-885.

Chavez JC, Agani F, Pichiule P, LaManna JC (2000b) Expression of hypoxia-inducible factor-1alpha in the brain of rats during chronic hypoxia. J Appl Physiol 89: 1937-1942.

Chavez JC, LaManna JC (2002) Activation of hypoxia-inducible factor-1 in the rat cerebral cortex after transient global ischemia: potential role of insulin-like growth factor-1. J Neurosci 22: 8922-8931.

Chikuma M, Masuda S, Kobayashi T, Nagao M, Sasaki R (2000) Tissue-specific regulation of erythropoietin production in the murine kidney, brain, and uterus. Am J Physiol Endocrinol Metab 279: E1242-E1248.

Dame C, Bartmann P, Wolber E, Fahnenstich H, Hofmann D, Fandrey J (2000) Erythropoietin gene expression in different areas of the developing human central nervous system. Brain Res Dev Brain Res 125: 69-74.

Digicaylioglu M, Bichet S, Marti HH, Wenger RH, Rivas LA, Bauer C, Gassmann M (1995) Localization of specific erythropoietin binding sites in defined areas of the mouse brain. Proc Natl Acad Sci U S A 92: 3717-3720.

Digicaylioglu M, Lipton SA (2001) Erythropoietin-mediated neuroprotection involves cross-talk between Jak2 and NF-kappaB signalling cascades. Nature 412: 641-647.

Ema M, Hirota K, Mimura J, Abe H, Yodoi J, Sogawa K, Poellinger L, Fujii-Kuriyama Y (1999) Molecular mechanisms of transcription activation by HLF and HIF1alpha in response to hypoxia: their stabilization and redox signal-induced interaction with CBP/p300. EMBO J 18: 1905-1914.

Epstein AC, Gleadle JM, McNeill LA, Hewitson KS, O'Rourke J, Mole DR, Mukherji M, Metzen E, Wilson MI, Dhanda A, Tian YM, Masson N, Hamilton DL, Jaakkola P, Barstead R, Hodgkin J, Maxwell PH, Pugh CW, Schofield CJ, Ratcliffe PJ (2001) C. elegans EGL-9 and mammalian homologs define a family of dioxygenases that regulate HIF by prolyl hydroxylation. Cell 107: 43-54.

Fandrey J, Bunn HF (1993) In vivo and in vitro regulation of erythropoietin mRNA: measurement by competitive polymerase chain reaction. Blood 81: 617-623.

Forsythe JA, Jiang BH, Iyer NV, Agani F, Leung SW, Koos RD, Semenza GL (1996) Activation of vascular endothelial growth factor gene transcription by hypoxia-inducible factor 1. Mol Cell Biol 16: 4604-4613.

Galson DL, Blanchard KL, Fandrey J, Goldberg MA, Bunn HF (1994) Cis elements that regulate the erythropoietin gene. Ann N Y Acad Sci 718: 21-30.

Gassmann M, Heinicke K, Soliz J, Ogunshola OO, Marti HH, Hofer T, Grimm C, Heinicke I, Egli B (2003) Non-erythroid functions of erythropoietin. Adv Exp Med Biol 543: 323-330.

Goldberg MA, Imagawa S, Strair RK, Bunn HF (1991) Regulation of the erythropoietin gene in Hep 3B cells. Semin Hematol 28: 35-40.

Gu J, Milligan J, Huang LE (2001) Molecular mechanism of hypoxia-inducible factor 1alpha -p300 interaction. A leucine-rich interface regulated by a single cysteine. J Biol Chem 276: 3550-3554.

Hirsila M, Koivunen P, Gunzler V, Kivirikko KI, Myllyharju J (2003) Characterization of the human prolyl 4-hydroxylases that modify the hypoxia-inducible factor. J Biol Chem 278: 30772-30780.

Ho V, Acquaviva A, Duh E, Bunn HF (1995) Use of a marked erythropoietin gene for investigation of its cis-acting elements. J Biol Chem 270: 10084-10090.

Huang J, Zhao Q, Mooney SM, Lee FS (2002) Sequence determinants in hypoxia-inducible factor-1alpha for hydroxylation by the prolyl hydroxylases PHD1, PHD2, and PHD3. J Biol Chem 277: 39792-39800.

Huang LE, Gu J, Schau M, Bunn HF (1998) Regulation of hypoxia-inducible factor 1alpha is mediated by an O2- dependent degradation domain via the ubiquitin-proteasome pathway. Proc Natl Acad Sci U S A 95: 7987-7992.

Imagawa S, Goldberg MA, Doweiko J, Bunn HF (1991) Regulatory elements of the erythropoietin gene. Blood 77: 278-285.

Ivan M, Haberberger T, Gervasi DC, Michelson KS, Gunzler V, Kondo K, Yang H, Sorokina I, Conaway RC, Conaway JW, Kaelin WG, Jr. (2002) Biochemical purification and pharmacological inhibition of a mammalian prolyl hydroxylase acting on hypoxia-inducible factor. Proc Natl Acad Sci U S A 99: 13459-13464.

Ivan M, Kondo K, Yang H, Kim W, Valiando J, Ohh M, Salic A, Asara JM, Lane WS, Kaelin WG, Jr. (2001) HIFalpha targeted for VHL-mediated destruction by proline hydroxylation: implications for O2 sensing. Science 292: 464-468.

Jaakkola P, Mole DR, Tian YM, Wilson MI, Gielbert J, Gaskell SJ, Kriegsheim A, Hebestreit HF, Mukherji M, Schofield CJ, Maxwell PH, Pugh CW, Ratcliffe PJ (2001) Targeting of HIF-alpha to the von Hippel-Lindau ubiquitylation complex by O2-regulated prolyl hydroxylation. Science 292: 468-472.

Jeong JW, Bae MK, Ahn MY, Kim SH, Sohn TK, Bae MH, Yoo MA, Song EJ, Lee KJ, Kim KW (2002) Regulation and destabilization of HIF-1alpha by ARD1-mediated acetylation. Cell 111: 709-720.

Jewell UR, Kvietikova I, Scheid A, Bauer C, Wenger RH, Gassmann M (2001) Induction of HIF-1alpha in response to hypoxia is instantaneous. FASEB J 15: 1312-1314.

Jiang BH, Zheng JZ, Leung SW, Roe R, Semenza GL (1997) Transactivation and inhibitory domains of hypoxia-inducible factor 1alpha. Modulation of transcriptional activity by oxygen tension. J Biol Chem 272: 19253-19260.

Juul SE, Anderson DK, Li Y, Christensen RD (1998) Erythropoietin and erythropoietin receptor in the developing human central nervous system. Pediatr Res 43: 40-49.

Katsura K, Kristian T, Siesjo BK (1994) Energy metabolism, ion homeostasis, and cell damage in the brain. Biochem Soc Trans 22: 991-996.

Kvietikova I, Wenger RH, Marti HH, Gassmann M (1995) The transcription factors ATF-1 and CREB-1 bind constitutively to the hypoxia-inducible factor-1 (HIF-1) DNA recognition site. Nucleic Acids Res 23: 4542-4550.

LaManna JC, Harik SI (1997) Brain metabolic and vascular adaptations to hypoxia in the rat. Review and update. Adv Exp Med Biol 428: 163-167.

LaManna JC, Vendel LM, Farrell RM (1992) Brain adaptation to chronic hypobaric hypoxia in rats. J Appl Physiol 72: 2238-2243.

Lando D, Peet DJ, Gorman JJ, Whelan DA, Whitelaw ML, Bruick RK (2002) FIH-1 is an asparaginyl hydroxylase enzyme that regulates the transcriptional activity of hypoxia-inducible factor. Genes Dev 16: 1466-1471.

Lopez-Lopez C, LeRoith D, Torres-Aleman I (2004) Insulin-like growth factor I is required for vessel remodeling in the adult brain. Proc Natl Acad Sci U S A 101: 9833-9838.

Mahon PC, Hirota K, Semenza GL (2001) FIH-1: a novel protein that interacts with HIF-1alpha and VHL to mediate repression of HIF-1 transcriptional activity. Genes Dev 15: 2675-2686.

Marti HH (1996) Erythropoietin gene expression in human, monkey and murine brain.

Marti HH, Gassmann M, Wenger RH, Kvietikova I, Morganti-Kossmann MC, Kossmann T, Trentz O, Bauer C (1997) Detection of erythropoietin in human liquor: intrinsic erythropoietin production in the brain. Kidney Int 51: 416-418.

Marti HH, Wenger RH, Rivas LA, Straumann U, Digicaylioglu M, Henn V, Yonekawa Y, Bauer C, Gassmann M (1996) Erythropoietin gene expression in human, monkey and murine brain. Eur J Neurosci 8: 666-676.

Masson N, Willam C, Maxwell PH, Pugh CW, Ratcliffe PJ (2001) Independent function of two destruction domains in hypoxia-inducible factor-alpha chains activated by prolyl hydroxylation. EMBO J 20: 5197-5206.

Masuda S, Chikuma M, Sasaki R (1997) Insulin-like growth factors and insulin stimulate erythropoietin production in primary cultured astrocytes. Brain Res 746: 63-70.

Masuda S, Okano M, Yamagishi K, Nagao M, Ueda M, Sasaki R (1994) A novel site of erythropoietin production. Oxygen-dependent production in cultured rat astrocytes. J Biol Chem 269: 19488-19493.

Maxwell PH, Pugh CW, Ratcliffe PJ (1993) Inducible operation of the erythropoietin 3' enhancer in multiple cell lines: evidence for a widespread oxygen-sensing mechanism. Proc Natl Acad Sci U S A 90: 2423-2427.

Maxwell PH, Wiesener MS, Chang GW, Clifford SC, Vaux EC, Cockman ME, Wykoff CC, Pugh CW, Maher ER, Ratcliffe PJ (1999) The tumour suppressor protein VHL targets hypoxia-inducible factors for oxygen-dependent proteolysis. Nature 399: 271-275.

Metzen E, Berchner-Pfannschmidt U, Stengel P, Marxsen JH, Stolze I, Klinger M, Huang WQ, Wotzlaw C, Hellwig-Burgel T, Jelkmann W, Acker H, Fandrey J (2003) Intracellular localisation of human HIF-1 alpha hydroxylases: implications for oxygen sensing. J Cell Sci 116: 1319-1326.

Monge C, Leon-Velarde F (1991) Physiological adaptation to high altitude: oxygen transport in mammals and birds. Physiol Rev 71: 1135-1172.

Nagai A, Nakagawa E, Choi HB, Hatori K, Kobayashi S, Kim SU (2001) Erythropoietin and erythropoietin receptors in human CNS neurons, astrocytes, microglia, and oligodendrocytes grown in culture. J Neuropathol Exp Neurol 60: 386-392.

Nakayama, K., Frew, IJ., Hagensen, M., Skals, M., Habelhah, H., Bhoumik, A., Kadoya, T., Erdjument-Bromage, H., Tempst, P., Frappell, PB., Bowdell, DD., and Ronai, Z. (2004) Siah2 regulates staibility of prolyl-hydroxylases, controls HIF1alpha abundance and modulates physiological resposnes to hypoxia. Cell 117(7): 941-52

Nielsen OJ, Schuster SJ, Kaufman R, Erslev AJ, Caro J (1987) Regulation of erythropoietin production in a human hepatoblastoma cell line. Blood 70: 1904-1909.

Pugh CW, Ebert BL, Ebrahim O, Maxwell PH, Ratcliffe PJ (1994a) Analysis of cis-acting sequences required for operation of the erythropoietin 3' enhancer in different cell lines. Ann N Y Acad Sci 718: 31-39.

Pugh CW, Ebert BL, Ebrahim O, Ratcliffe PJ (1994b) Characterisation of functional domains within the mouse erythropoietin 3' enhancer conveying oxygen-regulated responses in different cell lines. Biochim Biophys Acta 1217: 297-306.

Pugh CW, O'Rourke JF, Nagao M, Gleadle JM, Ratcliffe PJ (1997) Activation of hypoxia-inducible factor-1; definition of regulatory domains within the alpha subunit. J Biol Chem 272: 11205-11214.

Pugh CW, Tan CC, Jones RW, Ratcliffe PJ (1991) Functional analysis of an oxygen-regulated transcriptional enhancer lying 3' to the mouse erythropoietin gene. Proc Natl Acad Sci U S A 88: 10553-10557.

Ralph GS, Parham S, Lee SR, Beard GL, Craigon MH, Ward N, White JR, Barber RD, Rayner W, Kingsman SM, Mundy CR, Mazarakis ND, Krige D (2004) Identification of potential stroke targets by lentiviral vector mediated overexpression of HIF-1 alpha and HIF-2 alpha in a primary neuronal model of hypoxia. J Cereb Blood Flow Metab 24: 245-258.

Salceda S, Caro J (1997) Hypoxia-inducible factor 1alpha (HIF-1alpha) protein is rapidly degraded by the ubiquitin-proteasome system under normoxic conditions. Its stabilization by hypoxia depends on redox-induced changes. J Biol Chem 272: 22642-22647.

Semenza GL (1994a) Erythropoietin gene expression in transgenic mice and human hepatoma cells. Ann N Y Acad Sci 718: 41-47.

Semenza GL (1994b) Regulation of erythropoietin production. New insights into molecular mechanisms of oxygen homeostasis. Hematol Oncol Clin North Am 8: 863-884.

Semenza GL (1998) Hypoxia-inducible factor 1: master regulator of O2 homeostasis. Curr Opin Genet Dev 8: 588-594.

Semenza GL (1999) Regulation of mammalian O2 homeostasis by hypoxia-inducible factor 1. Annu Rev Cell Dev Biol 15: 551-578.

Semenza GL (2000a) HIF-1: mediator of physiological and pathophysiological responses to hypoxia. J Appl Physiol 88: 1474-1480.

Semenza GL (2000b) HIF-1: mediator of physiological and pathophysiological responses to hypoxia. J Appl Physiol 88: 1474-1480.

Semenza GL (2000c) Hypoxia, clonal selection, and the role of HIF-1 in tumor progression. Crit Rev Biochem Mol Biol 35: 71-103.

Semenza GL (2001) HIF-1, O(2), and the 3 PHDs: how animal cells signal hypoxia to the nucleus.

Semenza GL (2002) Physiology meets biophysics: visualizing the interaction of hypoxia-inducible factor 1 alpha with p300 and CBP. Proc Natl Acad Sci U S A 99: 11570-11572.

Semenza GL (2003a) Angiogenesis in ischemic and neoplastic disorders. Annu Rev Med 54: 17-28.

Semenza GL (2003b) Targeting HIF-1 for cancer therapy. Nat Rev Cancer 3: 721-732.

Semenza GL, Agani F, Booth G, Forsythe J, Iyer N, Jiang BH, Leung S, Roe R, Wiener C, Yu A (1997) Structural and functional analysis of hypoxia-inducible factor 1. Kidney Int 51: 553-555.

Semenza GL, Agani F, Feldser D, Iyer N, Kotch L, Laughner E, Yu A (2000) Hypoxia, HIF-1, and the pathophysiology of common human diseases. Adv Exp Med Biol 475: 123-130.

Semenza GL, Dureza RC, Traystman MD, Gearhart JD, Antonarakis SE (1990) Human erythropoietin gene expression in transgenic mice: multiple transcription initiation sites and cis-acting regulatory elements. Mol Cell Biol 10: 930-938.

Semenza GL, Koury ST, Nejfelt MK, Gearhart JD, Antonarakis SE (1991a) Cell-type-specific and hypoxia-inducible expression of the human erythropoietin gene in transgenic mice. Proc Natl Acad Sci U S A 88: 8725-8729.

Semenza GL, Nejfelt MK, Chi SM, Antonarakis SE (1991b) Hypoxia-inducible nuclear factors bind to an enhancer element located 3' to the human erythropoietin gene. Proc Natl Acad Sci U S A 88: 5680-5684.

Semenza GL, Roth PH, Fang HM, Wang GL (1994) Transcriptional regulation of genes encoding glycolytic enzymes by hypoxia-inducible factor 1. J Biol Chem 269: 23757-23763.

Semenza GL, Wang GL (1992) A nuclear factor induced by hypoxia via de novo protein synthesis binds to the human erythropoietin gene enhancer at a site required for transcriptional activation. Mol Cell Biol 12: 5447-5454.

Sharp FR, Bergeron M, Bernaudin M (2001) Hypoxia-inducible factor in brain. Adv Exp Med Biol 502: 273-291.

Sharp FR, Bernaudin M (2004) HIF1 and oxygen sensing in the brain. Nat Rev Neurosci 5: 437-448.

Siesjo BK (1981) Cell damage in the brain: a speculative synthesis. J Cereb Blood Flow Metab 1: 155-185.

Siesjo BK (1984) Cerebral circulation and metabolism. J Neurosurg 60: 883-908.

Siesjo BK, Plum F (1971) Cerebral energy metabolism in normoxia and in hypoxia. Acta Anaesthesiol Scand Suppl 45: 81-101.

Simon MC (2004) Siah proteins, HIF prolyl hydroxylases, and the physiological response to hypoxia. Cell 117: 851-853.

Siren AL, Knerlich F, Poser W, Gleiter CH, Bruck W, Ehrenreich H (2001) Erythropoietin and erythropoietin receptor in human ischemic/hypoxic brain. Acta Neuropathol (Berl) 101: 271-276.

Stolze I, Berchner-Pfannschmidt U, Freitag P, Wotzlaw C, Rossler J, Frede S, Acker H, Fandrey J (2002) Hypoxia-inducible erythropoietin gene expression in human neuroblastoma cells. Blood 100: 2623-2628.

Stroka DM, Burkhardt T, Desbaillets I, Wenger RH, Neil DA, Bauer C, Gassmann M, Candinas D (2001) HIF-1 is expressed in normoxic tissue and displays an organ-specific regulation under systemic hypoxia. FASEB J 15: 2445-2453.

Sugawa M, Sakurai Y, Ishikawa-Ieda Y, Suzuki H, Asou H (2002) Effects of erythropoietin on glial cell development; oligodendrocyte maturation and astrocyte proliferation. Neurosci Res 44: 391-403.

Talks KL, Turley H, Gatter KC, Maxwell PH, Pugh CW, Ratcliffe PJ, Harris AL (2000) The expression and distribution of the hypoxia-inducible factors HIF-1alpha and HIF-2alpha in normal human tissues, cancers, and tumor-associated macrophages. Am J Pathol 157: 411-421.

Tan CC, Eckardt KU, Ratcliffe PJ (1991) Organ distribution of erythropoietin messenger RNA in normal and uremic rats. Kidney Int 40: 69-76.

Tomita S, Ueno M, Sakamoto M, Kitahama Y, Ueki M, Maekawa N, Sakamoto H, Gassmann M, Kageyama R, Ueda N, Gonzalez FJ, Takahama Y (2003) Defective brain development in mice lacking the Hif-1alpha gene in neural cells. Mol Cell Biol 23: 6739-6749.

Wang GL, Jiang BH, Rue EA, Semenza GL (1995) Hypoxia-inducible factor 1 is a basic-helix-loop-helix-PAS heterodimer regulated by cellular O2 tension. Proc Natl Acad Sci U S A 92: 5510-5514.

Wang GL, Semenza GL (1993) General involvement of hypoxia-inducible factor 1 in transcriptional response to hypoxia. Proc Natl Acad Sci U S A 90: 4304-4308.

Wang GL, Semenza GL (1995) Purification and characterization of hypoxia-inducible factor 1. J Biol Chem 270: 1230-1237.

Wang GL, Semenza GL (1996) Molecular basis of hypoxia-induced erythropoietin expression. Curr Opin Hematol 3: 156-162.

Warnecke C, Griethe W, Weidemann A, Jurgensen JS, Willam C, Bachmann S, Ivashchenko Y, Wagner I, Frei U, Wiesener M, Eckardt KU (2003) Activation of the hypoxia-inducible factor-pathway and stimulation of angiogenesis by application of prolyl hydroxylase inhibitors. FASEB J 17: 1186-1188.

Wiesener MS, Jurgensen JS, Rosenberger C, Scholze CK, Horstrup JH, Warnecke C, Mandriota S, Bechmann I, Frei UA, Pugh CW, Ratcliffe PJ, Bachmann S, Maxwell PH, Eckardt KU (2003) Widespread hypoxia-inducible expression of HIF-2alpha in distinct cell populations of different organs. FASEB J 17: 271-273.

Yamaji R, Okada T, Moriya M, Naito M, Tsuruo T, Miyatake K, Nakano Y (1996) Brain capillary endothelial cells express two forms of erythropoietin receptor mRNA. Eur J Biochem 239: 494-500.

Yu AY, Shimoda LA, Iyer NV, Huso DL, Sun X, McWilliams R, Beaty T, Sham JS, Wiener CM, Sylvester JT, Semenza GL (1999) Impaired physiological responses to chronic hypoxia in mice partially deficient for hypoxia-inducible factor 1alpha. J Clin Invest 103: 691-696.

Yuan Y, Hilliard G, Ferguson T, Millhorn DE (2003) Cobalt inhibits the interaction between hypoxia-inducible factor-alpha and von Hippel-Lindau protein by direct binding to hypoxia-inducible factor-alpha. J Biol Chem 278: 15911-15916.

Zanjani ED, Ascensao JL, McGlave PB, Banisadre M, Ash RC (1981) Studies on the liver to kidney switch of erythropoietin production. J Clin Invest 67: 1183-1188.

Chapter 5

ERYTHROPOIETIN NEUROPROTECTION IN THE TERM AND PRETERM INFANT: SAFETY AND EFFICACY

Eric J. Demers, MD, and Sandra E. Juul, MD, PhD.
Department of Pediatrics, University of Washington, Seattle, Washington 98195

Abstract: The neonatal brain is particularly susceptible to a variety of insults, particularly hypoxia-ischemia (HI). Developmental differences occur during brain maturation and contribute to variable patterns of brain injury in preterm and term neonates after HI. Following HI, a multifactorial cascade is initiated that injures the developing brain and can lead to significant morbidity and mortality. Despite intense investigation, no effective interventions are currently available to lessen the devastating impact of neonatal brain injury. Erythropoietin (EPO), a hematologic cytokine, has shown promise as a neuroprotective agent in both *in vitro* and *in vivo* animal studies and continues to be actively investigated. In this chapter, we explore mechanisms of brain injury following HI in neonates, highlight differential injury responses that occur during development, and discuss the promising role of EPO as a therapeutic agent after neonatal HI.

Key words: Hypoxia-ischemia (HI), Brain Injury, Neonate, Preterm infant

1. INTRODUCTION

The developing human brain is vulnerable to a wide variety of genetic, developmental, and acquired abnormalities and insults. Brain injury can occur before, during, or after birth, and the timing of birth can range from extreme prematurity (22 to 23 weeks of gestation) to post term delivery (42 weeks). As the brain develops, there are selective vulnerabilities that occur at each developmental stage. The specific consequences of brain injury are therefore determined by the gestational age at which the injury occurred, and the severity of injury. Sequelae vary in type and severity, but generally involve one or more areas of development be it motor, cognitive, language, learning, or behavioral. Brain injury that occurs early in development may not become apparent until later, when developmental expectations (motor and cognitive) become more complex. In general, however, the earlier abnormal findings become apparent, the more severe the injury and its sequelae are likely to be.

There are currently no effective neuroprotective agents available for neonates; however, intense investigation is ongoing into several neuroprotective strategies. Erythropoietin (EPO), an endogenous hematopoietic cytokine, has both neurotrophic (Knabe et al., 2004) and neuroprotective effects (Juul, 2002). It is readily available in recombinant form (rEPO), and holds significant promise as a therapeutic agent for the treatment of a variety of adult and neonatal brain injuries. This chapter will discuss the developmental differences between newborn and adult brains, the mechanism(s) of brain injury associated with hypoxia-ischemia, EPO effects in the developing neonate, and the applicability of rEPO as a neuroprotective strategy in the setting of an injured, developing brain.

2. DEVELOPMENTAL DIFFERENCES BETWEEN PRETERM, NEONATAL AND MATURE BRAIN

There are developmental differences in injury susceptibility between the brains of preterm and term infants, and adults. In some aspects the immature brain is more susceptible to injury than the adult brain, while in others, it can be more resistant to injury. There are structural, biochemical, and cell-specific reasons for these differences in vulnerability. In addition, there are particular times at which the developing baby is at risk for injury: delivery and the immediate period afterwards is one particularly vulnerable period for both term and preterm infants. For preterm infants, the days immediately following birth are also particularly high risk for intracranial hemorrhage. By far the most common mechanism of brain injury in neonates is hypoxia-ischemia, and this will be the focus of discussion.

Hypoxia-ischemia injures the brain through a complex cascade of events that includes energy failure, ionic disruption, inflammation, and free radical generation, with subsequent cell death. Preterm and term neonates share several commonalities in these pathways, although development-specific susceptibilities are also present and may explain differences in the patterns of injury seen in preterm and term neonates. Similarities between preterm and term neonates in mechanism of injury mainly involve the generation of, and defense against free radical injury and the inflammatory cascade (Figure 1). Neonates are at increased risk of free radical attack due to high brain lipid content, a substrate for peroxidation (Hamrick and Ferriero, 2003). Their ability to defend against free radical attack is also diminished due to low glutathione peroxidase activity, decreased ability to upregulate catalase in response to injury (Volpe, 2001b), and low activity levels of Cu, Zn-superoxide dismutase (SOD) (Nishida et al., 1994). The importance of these antioxidants in neonatal brain injury is unclear, however, as mice overexpressing the antioxidant enzyme Cu, Zn-SOD had increased brain injury compared to control animals in a neonatal model of brain injury (Ditelberg et al., 1996), while in a similar model, mice overexpressing glutathione peroxidase had decreased injury (Sheldon et al., 2002). It is, perhaps, the balance of these anti-oxidant factors that is critical. Neonates have increased expression of both nNOS and iNOS during development, and their expression increases further with hypoxia-ischemia (Fernandez et al., 2003). The increased NOS activity is particularly notable in regions known to be glutamate receptor dense (Black et al., 1995), and increased expression of NO by the various NOS isoforms begins early after hypoxic-ischemic insult, persists for several days (Northington et al., 2001b) and contributes to free radical-induced membrane and DNA damage through production of the peroxynitrite radical.

The neonatal brain has more rapid microglial activation (Ivacko et al., 1996) and an increased expression of cytokines, particularly TNF-α and IL-6, both of which have been shown to be increased in neonates with brain injury (Yoon et al., 1997; Kadhim et al., 2003). In addition, neutrophils have been shown to contribute to brain injury in neonates (Hudome et al., 1997).

The developing brain has a high background rate of apoptosis normally, as there is an initial overproduction of neurons which then get culled: those neurons which have made effective synaptic connections are preserved, while cells that are not electrically active undergo apoptosis. This process is part of normal development, however, cells in the developing brain are also at increased risk to undergo apoptosis in response to injurious stimuli (Oppenheim, 1991; McDonald et al., 1997).

Factors which may protect the immature brain from injury include its lower rate of energy utilization, decreased basal oxygen consumption, increased capacity to maintain cellular energy through anaerobic metabolism, lower rate of accumulation of toxic by-products (particularly lactate), and the capability to utilize alternate energy sources, including lactate and ketone bodies (Gunn et al., 2001; Volpe, 2001a, c). The preterm brain has a cerebral metabolic rate for glucose that is approximately one third that of the adult brain (Powers et al., 1998). Despite these protective adaptations, the developing brain is at risk for injury, and this is further complicated by different cellular sensitivity during the course of development.

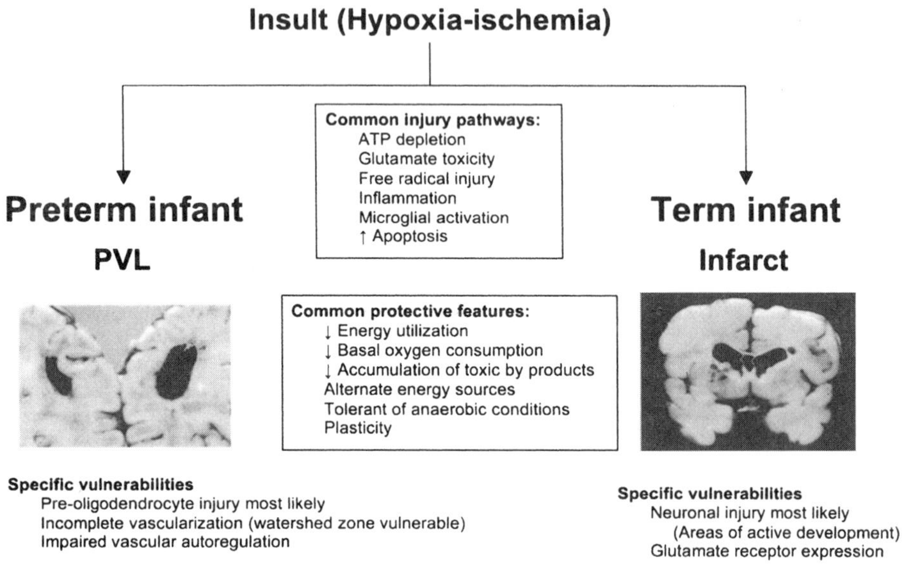

Figure 5-1. This figure illustrates commonalities in mechanisms of injury and protection between preterm and (left) and term (right) infant brains. Photographs of typical brain injuries for each gestational age are shown: PVL is the most common brain injury seen in preterm injury, while term infants frequently suffer from basal ganglia damage when exposed to significant hypoxia-ischemia. Photographs are shown courtesy of Drs. Raj Kapur, MD, PhD (Bilateral PVL, left worse than right) and Joseph R. Siebert, Ph.D. (cystic necrosis of the right putamen and subependymal germinolysis on contralateral side)

The most common brain injury affecting preterm infants is periventricular leukomalacia (PVL), a pattern of white matter injury (Volpe, 2001b). The preterm human brain at 24 to 32 weeks gestation is at highest risk for PVL, and this developmental window coincides with the presence of an oligodendrocyte precursor, the pre-oligodendrocyte, that is characterized immunochemically by reactivity to NG2 and O4, with no immunoreactivity to O1 antibody (Back et al., 1998; Back et al., 2001). During this period, the pre-oligodendrocyte is mitotically active and actively developing, and their healthy maturation and survival is influenced by both microglia and astrocytes (Pang et al., 2000). The pre-oligodendrocyte is quite sensitive to several insults, particularly free radical attack. The pre-oligodendrocyte can be injured by cystine deprivation, which decreases intracellular glutathione, and can be rescued by treatment with free radical scavengers (Yonezawa et al., 1996; Back et al., 1998). The pre-oligodendrocyte actively acquires iron, which is essential for subsequent myelin production. The high concentration of iron localized to these cells increases their risk of free radical injury due to the Fenton reaction (Hamrick and Ferriero, 2003).

Vascular development in the preterm brain is also immature. The preterm brain may have a pressure passive circulation with relatively low cerebral blood flow (Greisen, 1997), and a blunted vasodilatory response to hypoxia (Gunn et al., 2001) particularly in the white matter (Cavazzuti and Duffy, 1982). This combination results in the risk of hypoperfusion, particularly in less well vascularized watershed regions of the white matter (Lou et al., 1979a; Lou et al., 1979c; Lou et al., 1979b; Haruda, 2001; Volpe, 2001a; Hamrick and Ferriero, 2003). Both Boylan and Tsuji have shown that the impaired vascular autoregulation present in preterm neonates is highly associated with intraventricular hemorrhage or PVL (Boylan et al., 2000; Tsuji et al., 2000).

In contrast to preterm infants in which white matter injury predominates, brain injury in term infants affects primarily neurons (Volpe, 2001b). In the period immediately following injury, neurons in the thalamus, dentate gyrus, and habenula are at increased risk, while at later time points, neurons in the perirolandic cortex, basal ganglia (putamen), and CA 1 region of the hippocampus appear to be preferentially affected (Nakajima et al., 2000). This corresponds to regions of active neural development, particularly in the thalamus and basal ganglia (Hamrick and Ferriero, 2003). The term neonate is at particular risk of excitotoxic injury, especially that mediated through glutamate receptors, including both ionotropic (N-methyl-D-aspartate (NMDA) and α-amino-3-hydroxy-5-methyl-4-isoxazolepropionic acid (AMPA)) and metabotropic receptors. AMPA receptors increase over the course of gestation, and are expressed at different time points in oligodendrocytes and neurons (Archibald et al., 1999; Jensen, 2002). Similar developmental changes occur in NMDA receptor expression (Slater et al., 1993; Chahal et al., 1998). The increased susceptibility to excitotoxic injury is due in part to differences in glutamate receptor subtype expression, particularly the low expression of GluR2 in neonates compared to adults (Johnston, 2001). Neonatal glutamate receptors have a lower threshold for opening, a greater intracellular influx of Ca^{2+} compared to more mature receptors, and require a higher concentration of Mg^{2+} for antagonism (Monyer and Seeburg, 1993; Subramaniam and McGonigle, 1994; Chahal et al., 1998; Johnston, 2001). The combination of these factors leads to an increased susceptibility to seizures in the neonatal period that can be blocked by AMPA receptor antagonists (Jensen et al., 1995).

In summary, the neonatal brain is relatively resistant to energy depletion as compared to the adult brain. If however, a critical threshold of energy supply is exceeded, then the neonatal brain may be more susceptible to injury than the adult brain. Brain injury manifests itself differently between preterm and term infants due to differences in anatomy and cell susceptibility (oligodendrocytes are at high risk in preterm neonates, while neurons are more at risk in term neonates). Preterm and term infants share similar features that increase injury, including excitotoxic receptor expression, an underdeveloped antioxidant defense system and a heightened inflammatory response. In addition, the developing brain has a high background rate of apoptosis and this may be increased after an injurious stimulus. These factors are balanced by the neonatal brain's capacity to regenerate injured cellular elements or reorganize remaining elements to minimize deficits, a concept known as plasticity (Vaccarino and Ment, 2004). Plasticity is a feature of both the neonatal and adult brain and may be more robust in the developing brain (Murakami et al., 1992).

3. NEONATAL HYPOXIC-ISCHEMIC INJURY

Hypoxia-ischemia results in an acute energy crisis at the cellular level. The combined lack of oxygen and glucose interrupts oxidative phosphorylation, resulting in depleted energy stores (ATP). Although some irreversible damage occurs acutely during an asphyxial episode (primary energy failure), much of the damage to the neonatal brain occurs in the hours immediately following the episode, during the reperfusion period (secondary energy failure) (Vannucci, 1990; du Plessis and Johnston, 1997; Vannucci and Perlman, 1997). A cascade of biochemical events occurs during this period, including injury from free-radical formation, Ca^{++} accumulation, and neurotoxicity from glutamate, nitric oxide (NO), and a brisk inflammatory response (Vannucci, 1990). These events result in selective neuronal necrosis or apoptosis. Based on the postulate that one can intervene early in the course of apoptosis to reverse the process of cell death, and that this intervention would be of benefit, it is possible that an intervention initiated during the immediate post-injury period might be efficacious in reducing the severity of ongoing brain injury (Vannucci and Perlman, 1997).

In an elegant series of experiments, Northington et al. studied neuronal cell body, axonal, and terminal degeneration in brains from 7-day-old rat pups exposed to hypoxia-ischemia, at multiple time points ranging from 1.5 hours to 6 days after injury. They noted two phases of injury: early necrosis followed by later apoptosis, the specific timing of which varied by location in the brain (Northington et al., 2001b; Northington et al., 2001c; Northington et al., 2001a). Acute necrosis was present in the forebrain by three hours after injury, while delayed neurodegeneration became prominent in the thalamus at 24 hours (Northington et al., 2001b). At 48 hours after injury the ipsilateral cortex showed damage, and by six days the basal ganglia were affected in a secondary phase of injury. Factors that appeared to be active in this process of late neuronal death included Fas receptor stimulation, activation of caspases 3 and 8, and expression of pro-apoptotic Bcl-2 proteins such as Bax (Northington et al., 2001a).

Necrosis and apoptosis have long been conceived as two mutually exclusive pathways for cell death. Recently however, this view has been challenged, because within a given brain lesion (such as focal ischemia), single cells with a continuum of mixed morphological and biochemical characteristics of both apoptosis and necrosis can be identified. The mechanism of early cell death within the necrotic core of a lesion may be different from those activated during the secondary expansion of the lesion in the penumbral area, allowing for separate foci of potential intervention. Indeed, work by Benchoua et al. suggests that early cell death is dependent on pathways linked to caspase-8 and caspase-1, while cell death in the penumbral area includes caspase-9 pathways (Benchoua et al., 2001).

Important concepts underlying work to intervene in the cell injury and death that occurs following hypoxic-ischemic brain injury are first, that it is possible to do so, and second, that it is safe to do so. The safety of interfering with programmed cell death is particularly important when the patients are neonates, as apoptosis is an important part of normal brain development, and the proteins involved are precisely regulated (Hamner et al., 1999; Jarskog and Gilmore, 2000; Mooney and Miller, 2000).

4. MODELS OF HYPOXIC-ISCHEMIC BRAIN INJURY

Many models of neonatal hypoxic-ischemic brain injury are available, and each one has specific advantages and disadvantages. Details such as differences in brain structure, neurotransmitter metabolism, and the timing of development of specific receptors and enzyme activities, protein synthesis and gene expression must all be carefully considered when using animal models.

Rodent models are commonly used because they are inexpensive, easy to work with, and often respond to pharmacotherapy in a manner similar to humans. In addition, the use of knock out or knock down animals allow researchers to probe the specific role of individual proteins. One commonly used approach is the Rice-Vannucci model described in 1981 in which postnatal day 7 (P7) rats undergo unilateral common carotid artery ligation followed by variable periods of hypoxia (8% oxygen) (Rice et al., 1981).

How well the P7 rat brain corresponds to a newborn human brain varies with the particular focus of the investigator: the P7 rat brain is histologically similar to a 32 to 34 week human brain (Vannucci and Vannucci, 1997), the synapse formation, glutamic acid decarboxylase activity, choline acetyltransferase activity, and electrocortical function of a P10 to P12 day rat are more in keeping with the term human newborn (Romijn et al., 1991), while the growth/proliferation, persistence of a periventricular germinal matrix, neurochemical and metabolic data, EEG pattern, synapse formation, and patency of the blood–brain barrier of the P7-14 day rat correspond well to the term human newborn brain (Hagberg et al., 1997). Modifications of the Rice-Vannucci model have therefore included varying the postnatal age of the animal, the length of hypoxia exposure, the temperature at which the exposure occurs, and the application of the model to mice. Prolonged, hypoxic exposure alone has also been used to model perinatal brain injury (Stewart et al., 1997; Ment et al., 1998), as has, more recently, a model of transient ischemia (Ashwal et al., 1995). Although technically challenging due to the small size of the neonatal rat, this latter model incorporates reperfusion into the injury model, which is likely a very important component of naturally occurring hypoxic-ischemic injury in humans. Although rodent models can provide valuable insights regarding mechanisms of injury and repair, morphologic constraints (lissencephalic brain, high gray:white matter ratio), inability to combine in vivo neurologic testing with laboratory analysis, and limited scope of behavioral testing have made the translation of neuroprotective strategies from rodents to humans unsuccessful to date. To combat these problems, larger mammals including rabbits, puppies, piglets, sheep, and non-human primates have been used to model perinatal asphyxia (Stave, 1965; Myers, 1975; Ment et al., 1986; Vannucci, 1993; Adcock et al., 1996; Ment et al., 1997). Of the available models, the fetal and newborn rhesus monkey and immature rat have been studied most extensively because of their similarities to humans in respect to the physiology of reproduction and their neuroanatomy at or shortly following birth.

5. A NEW ROLE FOR EPO: NEUROPROTECTION

The study of EPO as a neuroprotective strategy is still relatively new. Most studies have been carried out in vitro, and in adult models of brain injury. Despite this, valuable information has been accumulated, much of which can be applied to the treatment of newborn infants.

5.1 Mechanisms of EPO action

The mechanism by which rEPO is neuroprotective remains an area of active investigation. Recombinant EPO has modest protective effects both in the central nervous system (Juul, 2002; Buemi et al., 2003) and the peripheral nervous system (Campana and Myers, 2001, 2003), and these effects are seen in models that employ a wide range of mechanisms of injury (Brines et al., 2000). This lack of specificity suggests that EPO may have multiple beneficial effects, some that affect neurons directly, and others that are likely secondary to EPO effects on other cells (Figure 2). To date, it has been shown that rEPO effects include: direct neurotrophic effects (Campana et al., 1998), decreased susceptibility to glutamate toxicity (Morishita et al., 1997; Kawakami et al., 2001), induction of anti-apoptotic factors (Juul et al., 1998b; Silva et al., 1999; Siren et al., 2001b; Celik et al., 2002; Renzi et al., 2002; Villa et al., 2003), decreased inflammation (Agnello et al., 2002; Gorio et al., 2002), decreased NO-mediated injury (Calapai et al., 2000; Digicaylioglu and Lipton, 2001; Kumral et al., 2004a), direct antioxidant effects (Chattopadhyay et al., 2000; Akisu et al., 2001a; Genc et al., 2002), and protective effects on glia (Nagai et al., 2001; Sugawa et al., 2002; Vairano et al., 2002). In vivo, EPO also increases erythropoiesis, which stimulates increased iron utilization. Although under normal circumstances free iron is scarce, in the face of hypoxic-ischemic injury, free iron is increased (Palmer et al., 1999). In the setting of hypoxia-ischemia, rEPO-induced increased iron utilization might decrease the availability of free iron, thereby preventing or decreasing free iron accumulation and its associated consequences (oxidative injury). Recombinant EPO may also provide neuroprotection by regulating blood flow to the brain following injury, as is suggested by rEPO neuroprotection in the model of subarachnoid hemorrhage (Grasso et al., 2002; Springborg et al., 2002). Recombinant EPO may also provide neuroprotection via its interaction with other growth factors such as vascular endothelial growth factor (VEGF) (Bocker-Meffert et al., 2002), and insulin growth factor-1 (IGF-1) (Chavez and LaManna, 2002). EPO has robust angiogenic effects (Jaquet et al., 2002), and works synergistically with VEGF (Ribatti et al., 1999). Increased angiogenesis during recuperation from a hypoxic injury might provide improved oxygen delivery to healing tissues (Siren et al., 2001a). Yet another possible mechanism of neuroprotection was proposed in a recent publication suggesting that rEPO decreases blood brain barrier permeability in the face of injury (Martinez-Estrada et al., 2003).

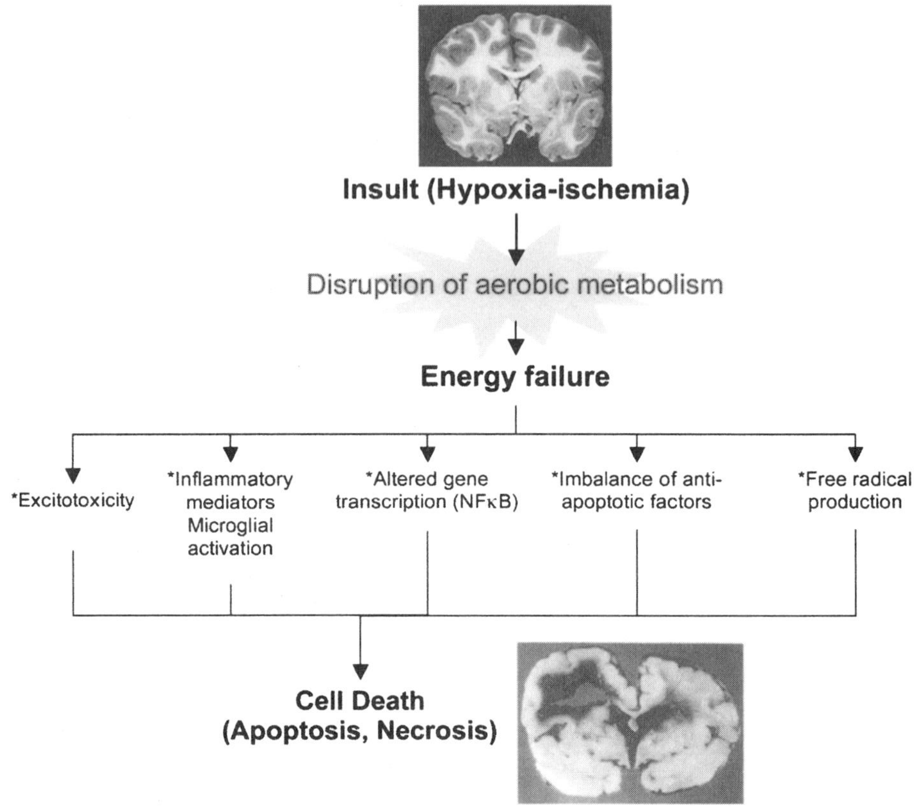

Figure 5-2. Sequence of events following hypoxia-ischemia in the central nervous system. * indicates possible points for rEPO to act. Photographs illustrating coronal sections of normal brain as compared to severely injured brain are shown courtesy of Dr. Raj Kapur, MD, PhD.

5.1.1 In vitro studies

EPO receptors are present on many cell types in the brain, and in vitro studies have been important in the determining how EPO might affect each of these cell types. These studies have demonstrated that in addition to the anti-apoptotic and neurotrophic effects EPO has on primary neurons and various neuronal cell lines (Juul et al., 1998b; Assandri et al., 1999; Koshimura et al., 1999; Kawakami et al., 2000; Lewczuk et al., 2000; Ruscher et al., 2002), rEPO also can influence astroglia (Sugawa et al., 2002), microglia (Chong et al., 2003), oligodendrocytes (Nagai et al., 2001; Sugawa et al., 2002), and retinal cell growth, differentiation and survival (Becerra and Amaral, 2002; Bocker-Meffert et al., 2002; Grimm et al., 2002;

Junk et al., 2002). Although important in establishing the spectrum of rEPO effects, these in vitro studies do not shed light on how rEPO might function differently in developing brain vs. mature brain, nor what the net effect of rEPO treatment might be in the more complex in vivo environment.

Neurons. Research using neuronal cell lines as well as primary cultures of neurons has clearly shown that rEPO is neuroprotective for cholinergic, dopaminergic, hippocampal, and cerebral cortical neurons. It can induce membrane depolarization, rapidly increase the cytosolic concentration of free calcium in neurons, increase dopamine synthesis and release, and increase cell viability in noxious conditions including the absence of serum or nerve growth factor, hypoxia, oxygen-glucose deprivation, or to NO or glutamate toxicity (Tabira et al., 1995; Morishita et al., 1997; Digicaylioglu and Lipton, 2001). This increased viability is dose-dependent, rEPO-specific, and calcium-mediated, as it is blunted by anti-EPO antibody and the calcium channel blocker nicardipine (Koshimura et al., 1999). RNA and protein synthesis are necessary for this protection, and maximal protection tends to occur when cells are pretreated, although protective effects can be seen up to six hours following injury (Morishita et al., 1997). There is some evidence that this neuroprotective function is mediated by a separate EPO-receptor (EPOR), which differs from the hematopoietic receptor in EPO affinity, size, and associated proteins (Masuda et al., 1993). A sequence of 17 amino acids within the EPO protein has been identified which exhibits neurotrophic, but not erythropoietic activity (Campana et al., 1998). This 17-mer, which acts by EPOR binding, induces neuronal differentiation, increases choline acetyltransferase activity and prevents cell death in two neuronal cell lines. Peptides such as this, or other EPO mimetic peptides (Wrighton et al., 1996) may provide another useful approach to neuroprotection, as their smaller size might improve blood brain barrier penetration. Other novel approaches include the use of modified rEPO molecules such as asialoerythropoietin, or carbamylated EPO. Asialoerythropoietin has improved blood brain barrier permeability when compared to rEPO, but has a short half life, and retains erythropoietic activity (Erbayraktar et al., 2003). In comparison, carbamylated EPO interacts specifically with the neuroprotective EPOR, has no erythropoietic effects, and appears to have similar blood brain barrier permeability as rEPO (Leist et al., 2004). These new approaches may allow for the use of smaller drug dosages and more specific neuroprotective activity, thus reducing potential toxicities.

Cultured neural stem cells have been used to determine the effects of EPO on neurogenesis. When these cells are exposed to hypoxia, EPO gene expression and neuronal number both increase. This enhanced neuron production is mimicked by rEPO stimulation, and blocked by co-administration of an EPO neutralizing antibody. This promotion of neurons occurs at the expense of multipotent progenitor cells. The developmental significance of this finding in vivo is unknown, but EPOR are present in the embryonic germinal zone, the subventricular zone, and the hippocampus (Juul et al., 1998a; Shingo et al., 2001), regions in which neurogenesis occurs, so this effect of EPO should be considered when investigating the safety and efficacy of rEPO treatment for preterm and term infants.

The cellular effects of rEPO vary depending on the cytokine milieu. For example, rEPO works synergistically with vascular endothelial growth factor (VEGF) to stimulate neovascularization (Ribatti et al., 1999). It has recently been shown that rEPO also works synergistically with insulin-like growth factor 1 (IGF-1) to protect cultured neurons from NMDA-mediated injury (Digicaylioglu et al., 2004). The mechanism of action involves the activation of phosphatidylinositol 3-kinase (PI3-K). Other environmental factors can also influence the effect of rEPO by changing receptor expression or sensitivity. For example, hypoxia and TNF-α upregulate EPOR expression by neurons, and hypoxia alone increases receptor sensitivity to rEPO (Chin et al., 2000; Nagai et al., 2001). Cell-specific differences in rEPO response, and even the cell signaling resulting from EPOR stimulation are emerging (Digicaylioglu and Lipton, 2001).

<u>Astrocytes.</u> Astrocytes are the primary source of endogenous EPO production in the brain (Masuda et al., 1994; Masuda et al., 1997), although neurons produce a lessor amount (Juul et al., 1998b; Bernaudin et al., 2000). Hypoxia (Masuda et al., 1994), insulin, and insulin-like growth factors (IGFs) (Masuda et al., 1997) stimulate EPO production by astrocytes in a dose-dependent manner. The effects of hypoxia and IGF-1 appear to be independent, as the stimulatory effect of IGF on EPO production is not affected by the oxygen concentration of astrocyte culture. New mRNA synthesis is required for EPO production, and the tyrosine kinase-signal transduction pathway is involved (Masuda et al., 1997; Bernaudin et al., 2000). Inflammatory cytokines IL-1 β, IL-6 and TNF-α downregulate EPO production by cultured astrocytes (Nagai et al., 2001).

Like neurons, astrocytes express EPOR, however, unlike neurons, EPO-stimulation does not protect astrocytes from hypoxic injury (Sinor and Greenberg, 2000; Vairano et al., 2002). This may be due to the absence of NFκB nuclear translocation with rEPO stimulation (Digicaylioglu and Lipton, 2001). Despite these differences in cell signaling and protective effects, rEPO (1, 3, 10 U/mL) enhances the proliferation of astrocytes (Sugawa et al., 2002).

<u>Microglia.</u> In primary cultures of rat cortical microglia and astrocytes, rEPO promotes microglial survival in a concentration-dependent manner. This effect is mediated via the Bcl-2 family, as rEPO shifts the Bcl-2:Bax

ratio towards a net anti-apoptotic effect (Vairano et al., 2002). In these experiments, rEPO stimulation did not affect the pro- inflammatory activity of microglial cells (Vairano et al., 2002). In contrast, in other studies, rEPO inhibited activation of cerebral microglial cells, and suppressed microglial phosphatidylserine receptor expression thereby decreasing phagocytosis (Chong et al., 2003; Villa et al., 2003). Thus the effects of EPO on microglia have not been fully clarified, so the impact of microglia on EPO-mediated neuroprotection is unresolved.

Oligodendrocytes. Oligodendrocytes are essential for white matter myelination, and injury to these cells may be an important factor in the development of PVL in preterm infants (Back et al., 2002). EPOR expression has been detected in O4-positive immature oligodendrocytes, and rEPO treatment or co-culture of these cells with astrocytes enhances oligodendrocyte maturation (Sugawa et al., 2002). This effect is inhibited by anti-EPO antibody and/or soluble EPOR, suggesting that release of EPO by astrocytes may promote oligodendrocyte differentiation. The consideration of rEPO as a candidate therapy for demyelinating diseases is a logical result of these experiments.

5.1.2 In vivo studies

Although endogenous EPO had been identified in the CSF of animal models and humans (Juul et al., 1997; Marti et al., 1997; Juul et al., 1999a), early studies evaluating the effects of rEPO in the central nervous system employed topical (intraventricular) applications of rEPO, as this large glycoprotein was not believed to cross the blood brain barrier (Juul et al., 1999a; Buemi et al., 2000). Indeed, at doses used to stimulate erythropoiesis (100-400 U/kg/dose intravenous [IV] or subcutaneous [SC]), this is likely true (Juul et al., 1999a; Buemi et al., 2000). However, it has now been shown that systemic administration of high dose rEPO (1,000 to 30,000 U/kg/dose intraperitoneal [IP]) can have significant neuroprotective effects (Brines et al., 2000; Kumral et al., 2004b). Although only a small proportion of rEPO crosses the intact blood brain barrier (Juul et al., 2004b), administration of high dose rEPO results in neuroprotective levels. Recombinant EPO concentration at the tissue level in brain is likely even higher in circumstances of injury, when the integrity of the blood brain barrier is breached. With only one exception (Wiessner et al., 2001), all publications to date have demonstrated that high dose rEPO treatment decreases both structural and behavioral abnormalities in adult models of central nervous system injury (Alafaci et al., 2000; Brines et al., 2000; Genc et al., 2001; Grasso, 2001; Agnello et al., 2002; Catania et al., 2002; Celik et al., 2002; Martinez-Estrada et al., 2003; Prass et al., 2003). Similar beneficial results have been noted in neonatal models of brain injury, although less data are available (Jones and Bergeron, 2001; Aydin et al., 2003; Kumral et al., 2003a; Solaroglu et al., 2003; Demers et al., 2004; Dzietko et al., 2004). Evaluation of the long-term safety of high dose rEPO treatment in neonates and adults is still in progress (Juul et al., 2004b; McPherson and Juul, 2004).

Neurodevelopment. EPO and its receptor are widely distributed in the developing embryo and fetus (Yasuda et al., 1993; Juul et al., 1998a), and their expression is tissue specific and developmentally controlled at the transcriptional level (Chikuma et al., 2000). High EPOR expression occurs in embryonic mouse neural tissues (Knabe et al., 2004), reaching levels comparable to adult bone marrow. This expression decreases by as much as 100- fold after birth (Liu et al., 1996; Liu et al., 1997; Moritz et al., 1997).

One proposed role for EPO is as a general morphogen and inducer of neurogenesis during early development (Yasuda et al., 1993; Wu et al., 1999; Ogilvie et al., 2000; Yu et al., 2002). EPOR and EPO knock out mice have been constructed, however, this mutation results in fetal demise at embryonic day 13.5 (Wu et al., 1995). Using tissues from these knock out embryos, it has been shown that in the absence of either EPO or the EPOR, there is a reduction in the number of neural progenitor cells, decreased neurogenesis, and increased neuronal apoptosis compared to wild type controls (Yu et al., 2002). Cultures of cortical neurons from EPOR knock out embryos also exhibit increased sensitivity to low oxygen tension (Yu et al., 2002). Other developmental abnormalities occur in these embryos, including ventricular hypoplasia, decreased cardiomyocyte proliferation and increased apoptosis in the liver. These defects can be corrected by insertion of the human EPO transgene (Wu et al., 1999; Yu et al., 2001).

Neuroprotection. Multiple animal models (mouse, rat, gerbil, rabbit, and pig) of adult brain injury have been used to investigate the effectiveness of rEPO as a neuroprotective agent (Brines et al., 2000; Marti et al., 2000; Siren et al., 2001b; Agnello et al., 2002; Catania et al., 2002; Celik et al., 2002; Romsi et al., 2002; Janjua and Mayer, 2003; Solaroglu et al., 2003). Mechanisms of brain injury have included focal and global cerebral ischemia, glutamate toxicity, spinal cord injury, subarachnoid hemorrhage, autoimmune encephalomyelitis and induction of seizures by kainic acid (Bernaudin et al., 1999; Brines et al., 2000; Springborg et al., 2002; Campana and Myers, 2003; Kaptanoglu et al., 2003; Sekiguchi et al., 2003). With such encouraging early results, rEPO investigations are being expanded to include Parkinson's disease, HIV mediated neuropathy, diabetic neuropathy, macular degeneration, retinitis pigmentosa, glaucoma, demyelinating injury, schizophrenia, and other types of neurodegenerative diseases (Becerra and Amaral, 2002; Lipton, 2004). Models of neonatal brain injury have also been used to determine whether rEPO effects established in adult models are present in the developing brain (Aydin et al., 2003; Kumral et al., 2003a; Demers et al., 2004; Dzietko et al., 2004; Kumral et al., 2004a; Kumral et al., 2004b; Wang et al., 2004). Despite the variability in approach (rEPO doses have ranged from 1,000-30,000 U/kg/dose IP with single or multiple doses), overall, rEPO has provided neuroprotection in neonatal models of injury that is commensurate with that shown in adult animal models. One novel approach has even included the injection of naked plasmid DNA encoding EPO to rescue animals exposed to hypoxia-ischemia (Wang et al., 2004). Optimal dosing regimens and duration of therapy are yet to be determined, and the long-term safety of high dose rEPO must be established. In addition, the applicability of rEPO to the white matter injury characteristic of preterm infants is yet to be established. Due to differences in pharmacokinetics between adults and neonates, and potential differences in blood brain barrier permeability, the dose required in neonates may be different than in adults (Ohls, 2002; Juul et al., 2004b). Furthermore, IV dosing may result in very different circulating concentrations of rEPO when compared to IP dosing, and this must be taken into consideration when applying the studies done in animal models to humans (Juul et al., 2004c).

6. IS INTERRUPTION OF APOPTOTIC CELL DEATH SAFE IN NEONATES?

In the developing mammalian brain, up to half of the neurons undergo apoptotic cell death. This occurs in most neuronal populations, including motor, sensory, autonomic neurons, and interneurons (Oppenheim, 1991). Apoptosis, or programmed cell death affects other brain cells as well, including oligodendrocytes (Barres et al., 1992). Two important families of proteins involved in the regulation of apoptosis are the Bcl and caspases. The Bcl family of proteins contains both anti-apoptotic (Bcl-2, Bcl-xL) and pro-apoptotic members (Bad, Bax, Bid). This family of proteins typically forms heterodimers between pro-apoptotic and anti-apoptotic members within the cytoplasm, and the balance between these opposing forces influences cell survival or death. In general, during periods of naturally occurring neuronal death, Bax expression is upregulated (pro-apoptotic), and the ratio of Bcl-2 (anti-apoptotic) to Bax expression decreases. This is associated with an increase in the expression of activated caspase 3. Reductions in caspase expression occur as the period of naturally occurring neuronal death subsides (Mooney and Miller, 2000). Moderation of apoptotic cell death is possible by influencing the balance of pro and anti-apoptotic gene expression. For example, using Bax knock out animals exposed to hypoxic-ischemic injury (unilateral carotid ligation followed by exposure to 45 minutes of 8% oxygen on postnatal day 7), Gibson et al. demonstrated 38% less hippocampal tissue loss when compared to wildtype mice (Gibson et al., 2001). Similarly, in experiments using targeted disruptions of members of the Bcl-2 and caspase gene families, reduced neuronal cell death occurs, resulting in increased numbers of neurons in a variety of neuronal subpopulations (Roth and D'Sa, 2001). These animals manifest specific, significant developmental delays (Rondi-Reig et al., 2001; Yan et al., 2001). These data suggest there is some risk to interfering with this normal developmental process, however, there may be a significant net benefit, despite the risk, in the face of hypoxia-ischemia.

7. OTHER EFFECTS OF EPO IN THE DEVELOPING NEONATE

EPOR are widely distributed in the developing human fetus. In addition to their presence in developing brain, they are present on vascular endothelium, hepatocytes, cardiomyocytes, intestinal enterocytes, among other cell types (Juul et al., 1998a). Specific effects of rEPO have only been studied in a few of these cell types.

7.1 Vascular effects of EPO

Functional EPOR are present on endothelial cells (Carlini et al., 1999; Ribatti et al., 1999; Jaquet et al., 2002). This is not surprising, as endothelial cells and primitive hematopoietic cells share a common embryonic origin, the hemangioblast (Choi, 1998; Choi et al., 1998; Robertson et al., 1999). The effects of rEPO on endothelial cells include: increased cell migration and proliferation, endothelin-1 production and release, enhanced vascular sensitivity to norepinephrine, and induction of angiogenesis (Bikfalvi and Han, 1994; Carlini et al., 1995b; Carlini et al., 1995a; Bode-Boger et al., 1996; Ribatti et al., 1999). The relationship between EPO, endothelin-1, and norepinephrine sensitivity may contribute to the hypertension seen in adult renal failure patients treated with rEPO. Although hypertension, and hypertensive leukoencephalopathy have been significant side effects of rEPO treatment in adults, hypertension is not a reported side effect of rEPO therapy in premature infants (Delanty et al., 1997).

Angiogenesis, the formation of new capillaries from pre-existing vessels, is a tightly controlled process essential for organogenesis during embryonic and fetal life. Under normal circumstances however, postnatally, the proliferation rate of endothelial cells is quite low. Pathologic exceptions include neovascularization during tumor progression, diabetic retinopathy and retinopathy of prematurity (ROP). During neovascularization, endothelial cells express an angiogenic phenotype which includes the production of proteases (which enhance the ability of cells to migrate), increased cell migration, and cell proliferation followed by redifferentiation (Risau, 1997). Stimulation of cultured endothelial cells with rEPO can induce this pro-angiogenic phenotype. Recombinant EPO also stimulates the formation of new blood vessels in the chick chorioallantoic membrane (Ribatti et al., 1999). Thus, rEPO alone can stimulate neovascularization both in vivo, and in vitro. EPO also interacts synergistically with VEGF, a potent angiogenic factor, to potentiate its vascular activity, and these two growth factors, both of which are regulated by hypoxia inducible factor (HIF-1α), often co-distribute (Hameed et al., 2004).

ROP is a serious complication of prematurity, occurring in approximately 30 to 40 percent of infants of birth weight $\leq$ 1500 g. There is an inverse relationship of ROP to birth weight. Five percent of infants $\leq$ 1500 g have severe ROP (grade 3+), often with disabling, life-long sequelae, including blindness, and there is some suggestion that the incidence may be increasing (Phelps, 1995; Hameed et al., 2004). These groups of very low birth weight infants are commonly treated with rEPO (100 to 400 U/kg/dose) for the treatment or prevention of anemia of prematurity. As research into rEPO-mediated neuroprotection progresses, these individuals might also be candidates for high dose rEPO therapy (1000-5000 U/kg/dose) to treat or prevent the white matter injury so common in this population. It is therefore of utmost importance to determine whether rEPO use in premature infants increases the risk of ROP. Although an association has been made between HIF-1 and ROP in rodent models (Morita et al., 2003), to date clinically, no increase in ROP has been associated with rEPO treatment, although studies have not been powered to look at this issue.

7.2 Gastrointestinal effects of EPO

Functional EPOR are present on enterocytes and endothelial cells in the developing small bowel (Juul et al., 1998a; Juul et al., 2001; Ashley et al., 2002). During development, these receptors are exposed systemically to circulating EPO, and topically to swallowed EPO in amniotic fluid (Teramo et al., 1987; Buescher et al., 1998). Postnatally, topical exposure may continue by ingestion of human milk (Kling et al., 1998; Juul et al., 2000). Although EPO is not essential for bowel growth and development (Yu et al., 2001; Juul et al., 2004a), it may contribute significantly to the normal growth and development of bowel during fetal and early postnatal life (Juul et al., 2001). In vitro effects of rEPO stimulation of enterocytes includes increasing cell migration, decreasing apoptotic cell death following injury, and trophic effects of gastric mucosal cells (Okada et al., 1996; Juul et al., 1999b). In vivo effects include improved bowel growth, improved wound healing, and possible protection from necrotizing enterocolitis (Fatouros et al., 1999; Ledbetter and Juul, 2000; Akisu et al., 2001b; Juul et al., 2001; Kumral et al., 2003b). If rEPO becomes a useful therapy for the treatment of brain injury in preterm infants, it might have additional beneficial effects in the developing intestinal tract.

8. CAVEAT FOR THE USE OF rEPO AS A NEUROTHERAPEUTIC AGENT IN NEWBORNS

Important issues unique to the developing brain must be addressed when targeting therapeutic strategies to the neurologically injured newborn. Safety and efficacy are two such issues. There is no question that decreasing injury from neonatal hypoxia-ischemia would be beneficial. Multiple strategies have been employed to date, but none have been successful. There is a strong rationale to enhance endogenously produced molecules after hypoxia-ischemia, as they are presumably safe and well tolerated. EPO is produced endogenously and may be such a compound. There are, however, potential concerns regarding rEPO usage, particularly at the higher dosing range required for neuroprotection. These include the production of hyperreactive platelets with subsequent thrombus formation (Wolf et al., 1997a; Wolf et al., 1997b), pure red cell aplasia secondary to anti-EPO antibodies (Casadevall, 2002; Indiveri and Murdaca, 2002; Zipursky, 2002; Casadevall, 2003) and hypertension (Vallance et al., 1988; Wong et al., 1990; Jabs and Harmon, 1996; Ismail and Ikizler, 1997). These complications have for the most part been reported in adult renal failure patients who receive rEPO over a prolonged period of time. Use of erythropoietic doses of rEPO (100 to 400 U/kg/dose IV or SQ) in neonatal populations has been very safe, with few side effects reported (Ohls, 1999). The long-term effects of high dose rEPO (1,000 to 30,000 U/kg/dose) in this population are unknown. Thus, although rEPO might effectively protect the adult brain from hypoxic-ischemic injury, this may not necessarily translate into safe and effective treatment for newborn brain injury. Because rEPO has potential widespread somatic and central nervous system activity, its use in newborns must be approached systematically and carefully. As noted previously, rEPO affects the growth and survival of many cell types in the brain, including neurons, glial cells, and vascular cells (Campana et al., 1998; Marti et al., 2000). EPO also has somatic effects, affecting the development of cardiovascular and intestinal structures (Wu et al., 1999; Juul et al., 2001). While some effects of rEPO may be critical to protecting the injured brain, it is still not known whether such therapy will have any negative or long-term impacts on neurodevelopment, or development of other systems. These issues are important, as it is now been shown that experiences which affect the fetus and newborn can influence health and well-being of the adult (Barker et al., 1989; Barker, 2000). There is great interest in Pediatrics to address problems such as perinatal asphyxia, PVL, and the sequelae of intraventricular hemorrhage. The initial studies of rEPO as a therapy for term and preterm infants with neurological injury are encouraging. The rodent models of unilateral hypoxic-ischemic injury show beneficial effects similar to those seen in adult models of injury (Kumral et al., 2003a; Demers et al., 2004; Kumral et al., 2004a). In addition to using neonatal rodent models of hypoxia-ischemia, our laboratory is also pursuing a primate model of perinatal asphyxia in which more sophisticated cognitive testing can be done, in addition to long-term safety studies. Before high dose rEPO is used for the treatment of brain injury in preterm and term infants, its long-term safety must be established, however, our hope is that rEPO may provide benefit in a field where there are few, if any, alternatives.

ACKNOWLEDGMENT

We are grateful to Drs Raj Kapur M.D., Ph.D. and Joseph R. Siebert, Ph.D. for their intellectual and photographic contributions to this chapter.

REFERENCES

Adcock LM, Yamashita Y, Goddard-Finegold J, Smith CV (1996) Cerebral hypoxia-ischemia increases microsomal iron in newborn piglets. Metab Brain Dis 11:359-367.

Agnello D, Bigini P, Villa P, Mennini T, Cerami A, Brines ML, Ghezzi P (2002) Erythropoietin exerts an anti-inflammatory effect on the CNS in a model of experimental autoimmune encephalomyelitis. Brain Res 952:128-134.

Akisu M, Tuzun S, Arslanoglu S, Yalaz M, Kultursay N (2001a) Effect of recombinant human erythropoietin administration on lipid peroxidation and antioxidant enzyme(s) activities in preterm infants. Acta Med Okayama 55:357-362.

Akisu M, Kullahcioglu Girgin F, Baka M, Husseyinov A, Kultursay N (2001b) The role of recombinant human erythropoietin in lipid peroxidation and platelet-activating factor generation in a rat model of necrotizing enterocolitis. Eur J Pediatr Surg 11:167-172.

Alafaci C, Salpietro F, Grasso G, Sfacteria A, Passalacqua M, Morabito A, Tripodo E, Calapai G, Buemi M, Tomasello F (2000) Effect of recombinant human erythropoietin on cerebral ischemia following experimental subarachnoid hemorrhage. Eur J Pharmacol 406:219-225.

Archibald K, Molnar E, Henley JM (1999) Differential changes in the subcellular distribution of alpha-amino-3-hydroxy-5-methyl-4-isoxazole propionate and N-methyl-D-aspartate receptors in neonate and adult rat cortex. Neurosci Lett 270:49-52.

Ashley RA, Dubuque SH, Dvorak B, Woodward SS, Williams SK, Kling PJ (2002) Erythropoietin stimulates vasculogenesis in neonatal rat mesenteric microvascular endothelial cells. Pediatr Res 51:472-478.

Ashwal S, Cole DJ, Osborne S, Osborne TN, Pearce WJ (1995) A new model of neonatal stroke: reversible middle cerebral artery occlusion in the rat pup. Pediatr Neurol 12:191-196.

Assandri R, Egger M, Gassmann M, Niggli E, Bauer C, Forster I, Gorlach A (1999) Erythropoietin modulates intracellular calcium in a human neuroblastoma cell line. J Physiol (Lond) 516:343-352.

Aydin A, Genc K, Akhisaroglu M, Yorukoglu K, Gokmen N, Gonullu E (2003) Erythropoietin exerts neuroprotective effect in neonatal rat model of hypoxic-ischemic brain injury. Brain Dev 25:494-498.

Back SA, Gan X, Li Y, Rosenberg PA, Volpe JJ (1998) Maturation-dependent vulnerability of oligodendrocytes to oxidative stress-induced death caused by glutathione depletion. J Neurosci 18:6241-6253.

Back SA, Luo NL, Borenstein NS, Volpe JJ, Kinney HC (2002) Arrested oligodendrocyte lineage progression during human cerebral white matter development: dissociation between the timing of progenitor differentiation and myelinogenesis. J Neuropathol Exp Neurol 61:197-211.

Back SA, Luo NL, Borenstein NS, Levine JM, Volpe JJ, Kinney HC (2001) Late oligodendrocyte progenitors coincide with the developmental window of vulnerability for human perinatal white matter injury. J Neurosci 21:1302-1312.

Barker DJ (2000) In utero programming of cardiovascular disease. Theriogenology 53:555-574.

Barker DJ, Winter PD, Osmond C, Margetts B, Simmonds SJ (1989) Weight in infancy and death from ischaemic heart disease. Lancet 2:577-580.

Barres BA, Hart IK, Coles HS, Burne JF, Voyvodic JT, Richardson WD, Raff MC (1992) Cell death in the oligodendrocyte lineage. J Neurobiol 23:1221-1230.

Becerra SP, Amaral J (2002) Erythropoietin--an endogenous retinal survival factor. N Engl J Med 347:1968-1970.

Benchoua A, Guegan C, Couriaud C, Hosseini H, Sampaio N, Morin D, Onteniente B (2001) Specific caspase pathways are activated in the two stages of cerebral infarction. J Neurosci 21:7127-7134.

Bernaudin M, Marti HH, Roussel S, Divoux D, Nouvelot A, MacKenzie ET, Petit E (1999) A potential role for erythropoietin in focal permanent cerebral ischemia in mice. J Cereb Blood Flow Metab 19:643-651.

Bernaudin M, Bellail A, Marti HH, Yvon A, Vivien D, Duchatelle I, Mackenzie ET, Petit E (2000) Neurons and astrocytes express EPO mRNA: oxygen-sensing mechanisms that involve the redox-state of the brain. Glia 30:271-278.

Bikfalvi A, Han ZC (1994) Angiogenic factors are hematopoietic growth factors and vice versa. Leukemia 8:523-529.

Black SM, Bedolli MA, Martinez S, Bristow JD, Ferriero DM, Soifer SJ (1995) Expression of neuronal nitric oxide synthase corresponds to regions of selective vulnerability to hypoxia-ischaemia in the developing rat brain. Neurobiol Dis 2:145-155.

Bocker-Meffert S, Rosenstiel P, Rohl C, Warneke N, Held-Feindt J, Sievers J, Lucius R (2002) Erythropoietin and VEGF promote neural outgrowth from retinal explants in postnatal rats. Invest Ophthalmol Vis Sci 43:2021-2026.

Bode-Boger SM, Boger RH, Kuhn M, Radermacher J, Frolich JC (1996) Recombinant human erythropoietin enhances vasoconstrictor tone via endothelin-1 and constrictor prostanoids. Kidney Int 50:1255-1261.

Boylan GB, Young K, Panerai RB, Rennie JM, Evans DH (2000) Dynamic cerebral autoregulation in sick newborn infants. Pediatr Res 48:12-17.

Brines ML, Ghezzi P, Keenan S, Agnello D, de Lanerolle NC, Cerami C, Itri LM, Cerami A (2000) Erythropoietin crosses the blood-brain barrier to protect against experimental brain injury. Proc Natl Acad Sci U S A 97:10526-10531.

Buemi M, Cavallaro E, Floccari F, Sturiale A, Aloisi C, Trimarchi M, Corica F, Frisina N (2003) The pleiotropic effects of erythropoietin in the central nervous system. J Neuropathol Exp Neurol 62:228-236.

Buemi M, Allegra A, Corica F, Floccari F, D'Avella D, Aloisi C, Calapai G, Iacopino G, Frisina N (2000) Intravenous recombinant erythropoietin does not lead to an increase in cerebrospinal fluid erythropoietin concentration. Nephrol Dial Transplant 15:422-423.

Buescher U, Hertwig K, Wolf C, Dudenhausen JW (1998) Erythropoietin in amniotic fluid as a marker of chronic fetal hypoxia. Int J Gynaecol Obstet 60:257-263.

Calapai G, Marciano MC, Corica F, Allegra A, Parisi A, Frisina N, Caputi AP, Buemi M (2000) Erythropoietin protects against brain ischemic injury by inhibition of nitric oxide formation. Eur J Pharmacol 401:349-356.

Campana WM, Myers RR (2001) Erythropoietin and erythropoietin receptors in the peripheral nervous system: changes after nerve injury. Faseb J 15:1804-1806.

Campana WM, Myers RR (2003) Exogenous erythropoietin protects against dorsal root ganglion apoptosis and pain following peripheral nerve injury. Eur J Neurosci 18:1497-1506.

Campana WM, Misasi R, O'Brien JS (1998) Identification of a neurotrophic sequence in erythropoietin. Int J Mol Med 1:235-241.

Carlini RG, Reyes AA, Rothstein M (1995a) Recombinant human erythropoietin stimulates angiogenesis in vitro. Kidney Int 47:740-745.

Carlini RG, Gupta A, Liapis H, Rothstein M (1995b) Endothelin-1 release by erythropoietin involves calcium signaling in endothelial cells. J Cardiovasc Pharmacol 26:889-892.

Carlini RG, Alonzo EJ, Dominguez J, Blanca I, Weisinger JR, Rothstein M, Bellorin-Font E (1999) Effect of recombinant human erythropoietin on endothelial cell apoptosis. Kidney Int 55:546-553.

Casadevall N (2002) Antibodies against rHuEPO: native and recombinant. Nephrol Dial Transplant 17 Suppl 5:42-47.

Casadevall N (2003) Pure red cell aplasia and anti-erythropoietin antibodies in patients treated with epoetin. Nephrol Dial Transplant 18 Suppl 8:viii37-41.

Catania MA, Marciano MC, Parisi A, Sturiale A, Buemi M, Grasso G, Squadrito F, Caputi AP, Calapai G (2002) Erythropoietin prevents cognition impairment induced by transient brain ischemia in gerbils. Eur J Pharmacol 437:147-150.

Cavazzuti M, Duffy TE (1982) Regulation of local cerebral blood flow in normal and hypoxic newborn dogs. Ann Neurol 11:247-257.

Celik M, Gokmen N, Erbayraktar S, Akhisaroglu M, Konakc S, Ulukus C, Genc S, Genc K, Sagiroglu E, Cerami A, Brines M (2002) Erythropoietin prevents motor neuron apoptosis and neurologic disability in experimental spinal cord ischemic injury. Proc Natl Acad Sci U S A 99:2258-2263.

Chahal H, D'Souza SW, Barson AJ, Slater P (1998) Modulation by magnesium of N-methyl-D-aspartate receptors in developing human brain. Arch Dis Child Fetal Neonatal Ed 78:F116-120.

Chattopadhyay A, Choudhury TD, Bandyopadhyay D, Datta AG (2000) Protective effect of erythropoietin on the oxidative damage of erythrocyte membrane by hydroxyl radical. Biochem Pharmacol 59:419-425.

Chavez JC, LaManna JC (2002) Activation of hypoxia-inducible factor-1 in the rat cerebral cortex after transient global ischemia: potential role of insulin-like growth factor-1. J Neurosci 22:8922-8931.

Chikuma M, Masuda S, Kobayashi T, Nagao M, Sasaki R (2000) Tissue-specific regulation of erythropoietin production in the murine kidney, brain, and uterus. Am J Physiol Endocrinol Metab 279:E1242-1248.

Chin K, Yu X, Beleslin-Cokic B, Liu C, Shen K, Mohrenweiser HW, Noguchi CT (2000) Production and processing of erythropoietin receptor transcripts in brain. Brain Res Mol Brain Res 81:29-42.

Choi K (1998) Hemangioblast development and regulation. Biochem Cell Biol 76:947-956.

Choi K, Kennedy M, Kazarov A, Papadimitriou JC, Keller G (1998) A common precursor for hematopoietic and endothelial cells. Development 125:725-732.

Chong ZZ, Kang JQ, Maiese K (2003) Erythropoietin fosters both intrinsic and extrinsic neuronal protection through modulation of microglia, Akt1, Bad, and caspase-mediated pathways. Br J Pharmacol 138:1107-1118.

Delanty N, Vaughan C, Frucht S, Stubgen P (1997) Erythropoietin-associated hypertensive posterior leukoencephalopathy. Neurology 49:686-689.

Demers E, McPherson RJ, Juul SE (2004) Erythropoietin ameliorates brain injury following hypoxia ischemia in neonatal rats. Pediatric Research In Press.

Digicaylioglu M, Lipton SA (2001) Erythropoietin-mediated neuroprotection involves cross-talk between Jak2 and NF-kappaB signalling cascades. Nature 412:641-647.

Digicaylioglu M, Garden G, Timberlake S, Fletcher L, Lipton SA (2004) Acute neuroprotective synergy of erythropoietin and insulin-like growth factor I. Proc Natl Acad Sci U S A 101:9855-9860.

Ditelberg JS, Sheldon RA, Epstein CJ, Ferriero DM (1996) Brain injury after perinatal hypoxia-ischemia is exacerbated in copper/zinc superoxide dismutase transgenic mice. Pediatr Res 39:204-208.

du Plessis AJ, Johnston MV (1997) Hypoxic-ischemic brain injury in the newborn: cellular mechanisms and potential strategies for neuroprotection. Clin Perinatol 24:627-654.

Dzietko M, Felderhoff-Mueser U, Sifringer M, Krutz B, Bittigau P, Thor F, Heumann R, Buhrer C, Ikonomidou C, Hansen HH (2004) Erythropoietin protects the developing brain against N-methyl-D-aspartate receptor antagonist neurotoxicity. Neurobiol Dis 15:177-187.

Erbayraktar S, Grasso G, Sfacteria A, Xie QW, Coleman T, Kreilgaard M, Torup L, Sager T, Erbayraktar Z, Gokmen N, Yilmaz O, Ghezzi P, Villa P, Fratelli M, Casagrande S, Leist M, Helboe L, Gerwein J, Christensen S, Geist MA, Pedersen LO, Cerami-Hand C, Wuerth JP, Cerami A, Brines M (2003) Asialoerythropoietin is a nonerythropoietic cytokine with broad neuroprotective activity in vivo. Proc Natl Acad Sci U S A 100:6741-6746.

Fatouros M, Dalekos GN, Mylonakis E, Vekinis G, Kappas AM (1999) Alterations in body weight, breaking strength, and wound healing in Wistar rats treated pre- and postoperatively with erythropoietin or granulocyte macrophage-colony stimulating factor: evidence of a previously unknown anabolic effect of erythropoietin? J Lab Clin Med 133:253-259.

Fernandez AP, Alonso D, Lisazoain I, Serrano J, Leza JC, Bentura ML, Lopez JC, Manuel Encinas J, Fernandez-Vizarra P, Castro-Blanco S, Martinez A, Martinez-Murillo R, Lorenzo P, Pedrosa JA, Peinado MA, Rodrigo J (2003) Postnatal changes in the nitric oxide system of the rat cerebral cortex after hypoxia during delivery. Brain Res Dev Brain Res 142:177-192.

Genc S, Akhisaroglu M, Kuralay F, Genc K (2002) Erythropoietin restores glutathione peroxidase activity in 1-methyl-4- phenyl-1,2,5,6-tetrahydropyridine-induced neurotoxicity in C57BL mice and stimulates murine astroglial glutathione peroxidase production in vitro. Neurosci Lett 321:73-76.

Genc S, Kuralay F, Genc K, Akhisaroglu M, Fadiloglu S, Yorukoglu K, Fadiloglu M, Gure A (2001) Erythropoietin exerts neuroprotection in 1-methyl-4-phenyl-1,2,3,6-tetrahydropyridine-treated C57/BL mice via increasing nitric oxide production. Neurosci Lett 298:139-141.

Gibson ME, Han BH, Choi J, Knudson CM, Korsmeyer SJ, Parsadanian M, Holtzman DM (2001) BAX contributes to apoptotic-like death following neonatal hypoxia-ischemia: evidence for distinct apoptosis pathways. Mol Med 7:644-655.

Gorio A, Gokmen N, Erbayraktar S, Yilmaz O, Madaschi L, Cichetti C, Di Giulio AM, Vardar E, Cerami A, Brines M (2002) Recombinant human erythropoietin counteracts secondary injury and markedly enhances neurological recovery from experimental spinal cord trauma. Proc Natl Acad Sci U S A 99:9450-9455.

Grasso G (2001) Neuroprotective effect of recombinant human erythropoietin in experimental subarachnoid hemorrhage. J Neurosurg Sci 45:7-14.

Grasso G, Buemi M, Alafaci C, Sfacteria A, Passalacqua M, Sturiale A, Calapai G, De Vico G, Piedimonte G, Salpietro FM, Tomasello F (2002) Beneficial effects of systemic administration of recombinant human erythropoietin in rabbits subjected to subarachnoid hemorrhage. Proc Natl Acad Sci U S A 99:5627-5631.

Greisen G (1997) Cerebral blood flow and energy metabolism in the newborn. Clin Perinatol 24:531-546.

Grimm C, Wenzel A, Groszer M, Mayser H, Seeliger M, Samardzija M, Bauer C, Gassmann M, Reme CE (2002) HIF-1-induced erythropoietin in the hypoxic retina protects against light-induced retinal degeneration. Nat Med 8:718-724.

Gunn AJ, Quaedackers JS, Guan J, Heineman E, Bennet L (2001) The premature fetus: not as defenseless as we thought, but still paradoxically vulnerable? Dev Neurosci 23:175-179.

Hagberg H, Bona E, Gilland E, Puka-Sundvall M (1997) Hypoxia-ischaemia model in the 7-day-old rat: possibilities and shortcomings. Acta Paediatr Suppl 422:85-88.

Hameed B, Shyamanur K, Kotecha S, Manktelow BN, Woodruff G, Draper ES, Field D (2004) Trends in the incidence of severe retinopathy of prematurity in a geographically defined population over a 10-year period. Pediatrics 113:1653-1657.

Hamner S, Skoglosa Y, Lindholm D (1999) Differential expression of bcl-w and bcl-x messenger RNA in the developing and adult rat nervous system. Neuroscience 91:673-684.

Hamrick SE, Ferriero DM (2003) The injury response in the term newborn brain: can we neuroprotect? Curr Opin Neurol 16:147-154.

Haruda FD (2001) The structure of blood vessels in the germinal matrix and the autoregulation of cerebral blood flow in premature infants. Pediatrics 108:1050-1051.

Hudome S, Palmer C, Roberts RL, Mauger D, Housman C, Towfighi J (1997) The role of neutrophils in the production of hypoxic-ischemic brain injury in the neonatal rat. Pediatr Res 41:607-616.

Indiveri F, Murdaca G (2002) Immunogenicity of erythropoietin and other growth factors. Rev Clin Exp Hematol Suppl 1:7-11.

Ismail N, Ikizler TA (1997) Erythropoietin-induced hypertension. J Med Liban 45:25-30.

Ivacko JA, Sun R, Silverstein FS (1996) Hypoxic-ischemic brain injury induces an acute microglial reaction in perinatal rats. Pediatr Res 39:39-47.

Jabs K, Harmon WE (1996) Recombinant human erythropoietin therapy in children on dialysis. Adv Ren Replace Ther 3:24-36.

Janjua N, Mayer SA (2003) Cerebral vasospasm after subarachnoid hemorrhage. Curr Opin Crit Care 9:113-119.

Jaquet K, Krause K, Tawakol-Khodai M, Geidel S, Kuck KH (2002) Erythropoietin and VEGF exhibit equal angiogenic potential. Microvasc Res 64:326-333.

Jarskog LF, Gilmore JH (2000) Developmental expression of Bcl-2 protein in human cortex. Brain Res Dev Brain Res 119:225-230.

Jensen FE (2002) The role of glutamate receptor maturation in perinatal seizures and brain injury. Int J Dev Neurosci 20:339-347.

Jensen FE, Blume H, Alvarado S, Firkusny I, Geary C (1995) NBQX blocks acute and late epileptogenic effects of perinatal hypoxia. Epilepsia 36:966-972.

Johnston MV (2001) Excitotoxicity in neonatal hypoxia. Ment Retard Dev Disabil Res Rev 7:229-234.

Jones NM, Bergeron M (2001) Hypoxic preconditioning induces changes in HIF-1 target genes in neonatal rat brain. J Cereb Blood Flow Metab 21:1105-1114.

Junk AK, Mammis A, Savitz SI, Singh M, Roth S, Malhotra S, Rosenbaum PS, Cerami A, Brines M, Rosenbaum DM (2002) Erythropoietin administration protects retinal neurons from acute ischemia-reperfusion injury. Proc Natl Acad Sci U S A 99:10659-10664.

Juul S (2002) Erythropoietin in the central nervous system, and its use to prevent hypoxic-ischemic brain damage. Acta Paediatr Suppl 91:36-42.

Juul SE, Yachnis AT, Christensen RD (1998a) Tissue distribution of erythropoietin and erythropoietin receptor in the developing human fetus. Early Hum Dev 52:235-249.

Juul SE, Stallings SA, Christensen RD (1999a) Erythropoietin in the cerebrospinal fluid of neonates who sustained CNS injury. Pediatr Res 46:543-547.

Juul SE, McPherson RJ, Kapur RP (2004a) Erythropoietin receptor function is not required for normal bowel development. Pediatr Res 55:522A.

Juul SE, Harcum J, Li Y, Christensen RD (1997) Erythropoietin is present in the cerebrospinal fluid of neonates. J Pediatr 130:428-430.

Juul SE, Anderson DK, Li Y, Christensen RD (1998b) Erythropoietin and erythropoietin receptor in the developing human central nervous system. Pediatr Res 43:40-49.

Juul SE, Joyce AE, Zhao Y, Ledbetter DJ (1999b) Why is erythropoietin present in human milk? Studies of erythropoietin receptors on enterocytes of human and rat neonates. Pediatr Res 46:263-268.

Juul SE, Zhao Y, Dame JB, Du Y, Hutson AD, Christensen RD (2000) Origin and fate of erythropoietin in human milk. Pediatr Res 48:660-667.

Juul SE, McPherson RJ, Farrell FX, Jolliffe L, Ness DJ, Gleason CA (2004b) Erythropoietin Concentrations in Cerebrospinal Fluid of Nonhuman Primates and Fetal Sheep following High-Dose Recombinant Erythropoietin. Biol Neonate 85:138-144.

Juul SE, McPherson RJ, Farrell FX, Jolliffe L, Ness DJ, Gleason CA (2004c) Erytropoietin concentrations in cerebrospinal fluid of nonhuman primates and fetal sheep following high-dose recombinant erythropoietin. Biol Neonate 85:138-144.

Juul SE, Ledbetter DJ, Joyce AE, Dame C, Christensen RD, Zhao Y, DeMarco V (2001) Erythropoietin acts as a trophic factor in neonatal rat intestine. Gut 49:182-189.

Kadhim H, Tabarki B, De Prez C, Sebire G (2003) Cytokine immunoreactivity in cortical and subcortical neurons in periventricular leukomalacia: are cytokines implicated in neuronal dysfunction in cerebral palsy? Acta Neuropathol (Berl) 105:209-216.

Kaptanoglu E, Solaroglu I, Okutan O, Surucu HS, Akbiyik F, Beskonakli E (2003) Erythropoietin exerts neuroprotection after acute spinal cord injury in rats: effect on lipid peroxidation and early ultrastructural findings. Neurosurg Rev.

Kawakami M, Iwasaki S, Sato K, Takahashi M (2000) Erythropoietin inhibits calcium-induced neurotransmitter release from clonal neuronal cells. Biochem Biophys Res Commun 279:293-297.

Kawakami M, Sekiguchi M, Sato K, Kozaki S, Takahashi M (2001) Erythropoietin receptor-mediated inhibition of exocytotic glutamate release confers neuroprotection during chemical ischemia. J Biol Chem 276:39469-39475.

Kling PJ, Sullivan TM, Roberts RA, Philipps AF, Koldovsky O (1998) Human milk as a potential enteral source of erythropoietin. Pediatr Res 43:216-221.

Knabe W, Knerlich F, Washausen S, Kietzmann T, Siren AL, Brunnett G, Kuhn HJ, Ehrenreich H (2004) Expression patterns of erythropoietin and its receptor in the developing midbrain. Anat Embryol (Berl) 207:503-512.

Koshimura K, Murakami Y, Sohmiya M, Tanaka J, Kato Y (1999) Effects of erythropoietin on neuronal activity. J Neurochem 72:2565-2572.

Kumral A, Ozer E, Yilmaz O, Akhisaroglu M, Gokmen N, Duman N, Ulukus C, Genc S, Ozkan H (2003a) Neuroprotective effect of erythropoietin on hypoxic-ischemic brain injury in neonatal rats. Biol Neonate 83:224-228.

Kumral A, Baskin H, Duman N, Yilmaz O, Tatli M, Ozer E, Gokmen N, Genc S, Ozkan H (2003b) Erythropoietin protects against necrotizing enterocolitis of newborn rats by the inhibiting nitric oxide formation. Biol Neonate 84:325-329.

Kumral A, Baskin H, Gokmen N, Yilmaz O, Genc K, Genc S, Tatli MM, Duman N, Ozer E, Ozkan H (2004a) Selective Inhibition of Nitric Oxide in Hypoxic-Ischemic Brain Model in Newborn Rats: Is It an Explanation for the Protective Role of Erythropoietin? Biol Neonate 85:51-54.

Kumral A, Uysal N, Tugyan K, Sonmez A, Yilmaz O, Gokmen N, Kiray M, Genc S, Duman N, Koroglu TF, Ozkan H, Genc K (2004b) Erythropoietin improves long-term spatial memory deficits and brain injury following neonatal hypoxia-ischemia in rats. Behav Brain Res 153:77-86.

Ledbetter DJ, Juul SE (2000) Erythropoietin and the incidence of necrotizing enterocolitis in infants with very low birth weight. J Pediatr Surg 35:178-181.

Leist M, Ghezzi P, Grasso G, Bianchi R, Villa P, Fratelli M, Savino C, Bianchi M, Nielsen J, Gerwien J, Kallunki P, Larsen AK, Helboe L, Christensen S, Pedersen LO, Nielsen M, Torup L, Sager T, Sfacteria A, Erbayraktar S, Erbayraktar Z, Gokmen N, Yilmaz O, Cerami-Hand C, Xie QW, Coleman T, Cerami A, Brines M (2004) Derivatives of erythropoietin that are tissue protective but not erythropoietic. Science 305:239-242.

Lewczuk P, Hasselblatt M, Kamrowski-Kruck H, Heyer A, Unzicker C, Siren AL, Ehrenreich H (2000) Survival of hippocampal neurons in culture upon hypoxia: effect of erythropoietin. Neuroreport 11:3485-3488.

Lipton SA (2004) Erythropoietin for neurologic protection and diabetic neuropathy. N Engl J Med 350:2516-2517.

Liu C, Shen K, Liu Z, Noguchi CT (1997) Regulated human erythropoietin receptor expression in mouse brain. J Biol Chem 272:32395-32400.

Liu C, Yu K, Shen K, Liu Z, Noguchi CT (1996) Transgenic mice containing the human erythropoietin receptor gene exhibit correct hematopoietic and neural expression. Proc Assoc Am Physicians 108:449-454.

Lou HC, Lassen NA, Friis-Hansen B (1979a) Impaired autoregulation of cerebral blood flow in the distressed newborn infant. J Pediatr 94:118-121.

Lou HC, Skov H, Pedersen H (1979b) Low cerebral blood flow: a risk factor in the neonate. J Pediatr 95:606-609.

Lou HC, Lassen NA, Tweed WA, Johnson G, Jones M, Palahniuk RJ (1979c) Pressure passive cerebral blood flow and breakdown of the blood-brain barrier in experimental fetal asphyxia. Acta Paediatr Scand 68:57-63.

Marti HH, Bernaudin M, Petit E, Bauer C (2000) Neuroprotection and angiogenesis: dual role of erythropoietin in brain ischemia. News Physiol Sci 15:225-229.

Marti HH, Gassmann M, Wenger RH, Kvietikova I, Morganti-Kossmann MC, Kossmann T, Trentz O, Bauer C (1997) Detection of erythropoietin in human liquor: intrinsic erythropoietin production in the brain. Kidney Int 51:416-418.

Martinez-Estrada OM, Rodriguez-Millan E, Gonzalez-De Vicente E, Reina M, Vilaro S, Fabre M (2003) Erythropoietin protects the in vitro blood-brain barrier against VEGF-induced permeability. Eur J Neurosci 18:2538-2544.

Masuda S, Chikuma M, Sasaki R (1997) Insulin-like growth factors and insulin stimulate erythropoietin production in primary cultured astrocytes. Brain Res 746:63-70.

Masuda S, Okano M, Yamagishi K, Nagao M, Ueda M, Sasaki R (1994) A novel site of erythropoietin production. Oxygen-dependent production in cultured rat astrocytes. J Biol Chem 269:19488-19493.

Masuda S, Nagao M, Takahata K, Konishi Y, Gallyas F, Jr., Tabira T, Sasaki R (1993) Functional erythropoietin receptor of the cells with neural characteristics. Comparison with receptor properties of erythroid cells. J Biol Chem 268:11208-11216.

McDonald JW, Behrens MI, Chung C, Bhattacharyya T, Choi DW (1997) Susceptibility to apoptosis is enhanced in immature cortical neurons. Brain Res 759:228-232.

McPherson RJ, Juul SE (2004) Treatment of Neonatal Rats with High Dose Erythropoietin (Epo) Produces Lasting Effects. Pediatric Research In Press.

Ment LR, Schwartz M, Makuch RW, Stewart WB (1998) Association of chronic sublethal hypoxia with ventriculomegaly in the developing rat brain. Brain Res Dev Brain Res 111:197-203.

Ment LR, Stewart WB, Duncan CC, Pitt BR, Cole JS (1986) Beagle puppy model of perinatal cerebral infarction. Regional cerebral prostaglandin changes during acute hypoxemia. J Neurosurg 65:851-855.

Ment LR, Stewart WB, Fronc R, Seashore C, Mahooti S, Scaramuzzino D, Madri JA (1997) Vascular endothelial growth factor mediates reactive angiogenesis in the postnatal developing brain. Brain Res Dev Brain Res 100:52-61.

Monyer H, Seeburg PH (1993) Constituents involved in glutamate receptor signaling. Hippocampus 3 Spec No:125-129.

Mooney SM, Miller MW (2000) Expression of bcl-2, bax, and caspase-3 in the brain of the developing rat. Brain Res Dev Brain Res 123:103-117.

Morishita E, Masuda S, Nagao M, Yasuda Y, Sasaki R (1997) Erythropoietin receptor is expressed in rat hippocampal and cerebral cortical neurons, and erythropoietin prevents in vitro glutamate- induced neuronal death. Neuroscience 76:105-116.

Morita M, Ohneda O, Yamashita T, Takahashi S, Suzuki N, Nakajima O, Kawauchi S, Ema M, Shibahara S, Udono T, Tomita K, Tamai M, Sogawa K, Yamamoto M, Fujii-Kuriyama

Y (2003) HLF/HIF-2alpha is a key factor in retinopathy of prematurity in association with erythropoietin. Embo J 22:1134-1146.

Moritz KM, Lim GB, Wintour EM (1997) Developmental regulation of erythropoietin and erythropoiesis. Am J Physiol 273:R1829-1844.

Murakami F, Song WJ, Katsumaru H (1992) Plasticity of neuronal connections in developing brains of mammals. Neurosci Res 15:235-253.

Myers RE (1975) Four patterns of perinatal brain damage and their conditions of occurrence in primates. Adv Neurol 10:223-234.

Nagai A, Nakagawa E, Choi HB, Hatori K, Kobayashi S, Kim SU (2001) Erythropoietin and erythropoietin receptors in human CNS neurons, astrocytes, microglia, and oligodendrocytes grown in culture. J Neuropathol Exp Neurol 60:386-392.

Nakajima W, Ishida A, Lange MS, Gabrielson KL, Wilson MA, Martin LJ, Blue ME, Johnston MV (2000) Apoptosis has a prolonged role in the neurodegeneration after hypoxic ischemia in the newborn rat. J Neurosci 20:7994-8004.

Nishida A, Misaki Y, Kuruta H, Takashima S (1994) Developmental expression of copper, zinc-superoxide dismutase in human brain by chemiluminescence. Brain Dev 16:40-43.

Northington FJ, Ferriero DM, Martin LJ (2001a) Neurodegeneration in the thalamus following neonatal hypoxia-ischemia is programmed cell death. Dev Neurosci 23:186-191.

Northington FJ, Ferriero DM, Flock DL, Martin LJ (2001b) Delayed neurodegeneration in neonatal rat thalamus after hypoxia- ischemia is apoptosis. J Neurosci 21:1931-1938.

Northington FJ, Ferriero DM, Graham EM, Traystman RJ, Martin LJ (2001c) Early Neurodegeneration after Hypoxia-Ischemia in Neonatal Rat Is Necrosis while Delayed Neuronal Death Is Apoptosis. Neurobiol Dis 8:207-219.

Ogilvie M, Yu X, Nicolas-Metral V, Pulido SM, Liu C, Ruegg UT, Noguchi CT (2000) Erythropoietin stimulates proliferation and interferes with differentiation of myoblasts. J Biol Chem 275:39754-39761.

Ohls RK (1999) Erythropoietin to prevent and treat the anemia of prematurity. Curr Opin Pediatr 11:108-114.

Ohls RK (2002) Human recombinant erythropoietin in the prevention and treatment of anemia of prematurity. Paediatr Drugs 4:111-121.

Okada A, Kinoshita Y, Maekawa T, Hassan MS, Kawanami C, Asahara M, Matsushima Y, Kishi K, Nakata H, Naribayashi Y, Chiba T (1996) Erythropoietin stimulates proliferation of rat-cultured gastric mucosal cells. Digestion 57:328-332.

Oppenheim RW (1991) Cell death during development of the nervous system. Annu Rev Neurosci 14:453-501.

Palmer C, Menzies SL, Roberts RL, Pavlick G, Connor JR (1999) Changes in iron histochemistry after hypoxic-ischemic brain injury in the neonatal rat. J Neurosci Res 56:60-71.

Pang Y, Cai Z, Rhodes PG (2000) Effects of lipopolysaccharide on oligodendrocyte progenitor cells are mediated by astrocytes and microglia. J Neurosci Res 62:510-520.

Phelps DL (1995) Retinopathy of prematurity. Pediatr Rev 16:50-56.

Powers WJ, Rosenbaum JL, Dence CS, Markham J, Videen TO (1998) Cerebral glucose transport and metabolism in preterm human infants. J Cereb Blood Flow Metab 18:632-638.

Prass K, Scharff A, Ruscher K, Lowl D, Muselmann C, Victorov I, Kapinya K, Dirnagl U, Meisel A (2003) Hypoxia-induced stroke tolerance in the mouse is mediated by erythropoietin. Stroke 34:1981-1986.

Renzi MJ, Farrell FX, Bittner A, Galindo JE, Morton M, Trinh H, Jolliffe LK (2002) Erythropoietin induces changes in gene expression in PC-12 cells. Brain Res Mol Brain Res 104:86-95.

Ribatti D, Presta M, Vacca A, Ria R, Giuliani R, Dell'Era P, Nico B, Roncali L, Dammacco F (1999) Human erythropoietin induces a pro-angiogenic phenotype in cultured endothelial cells and stimulates neovascularization In vivo. Blood 93:2627-2636.

Rice JE, 3rd, Vannucci RC, Brierley JB (1981) The influence of immaturity on hypoxic-ischemic brain damage in the rat. Ann Neurol 9:131-141.

Risau W (1997) Mechanisms of angiogenesis. Nature 386:671-674.

Robertson S, Kennedy M, Keller G (1999) Hematopoietic commitment during embryogenesis. Ann N Y Acad Sci 872:9-15; discussion 15-16.

Romijn HJ, Hofman MA, Gramsbergen A (1991) At what age is the developing cerebral cortex of the rat comparable to that of the full-term newborn human baby? Early Hum Dev 26:61-67.

Romsi P, Ronka E, Kiviluoma K, Vainionpaa V, Hirvonen J, Mennander A, Pokela M, Biancari F, Rimpilainen J, Juvonen T (2002) Potential neuroprotective benefits of erythropoietin during experimental hypothermic circulatory arrest. J Thorac Cardiovasc Surg 124:714-723.

Rondi-Reig L, Lemaigre-Dubreuil Y, Montecot C, Muller D, Martinou JC, Caston J, Mariani J (2001) Transgenic mice with neuronal overexpression of bcl-2 gene present navigation disabilities in a water task. Neuroscience 104:207-215.

Roth KA, D'Sa C (2001) Apoptosis and brain development. Ment Retard Dev Disabil Res Rev 7:261-266.

Ruscher K, Freyer D, Karsch M, Isaev N, Megow D, Sawitzki B, Priller J, Dirnagl U, Meisel A (2002) Erythropoietin is a paracrine mediator of ischemic tolerance in the brain: evidence from an in vitro model. J Neurosci 22:10291-10301.

Sekiguchi Y, Kikuchi S, Myers RR, Marie Campana W (2003) ISSLS Prize Winner: Erythropoietin Inhibits Spinal Neuronal Apoptosis and Pain Following Nerve Root Crush. Spine 28:2577-2584.

Sheldon RA, Almli L, Ferriero DM (2002) Copper/zinc superoxide dismutase transgenic brain in neonatal hypoxia-ischemia. Methods Enzymol 353:389-397.

Shingo T, Sorokan ST, Shimazaki T, Weiss S (2001) Erythropoietin regulates the in vitro and in vivo production of neuronal progenitors by mammalian forebrain neural stem cells. J Neurosci 21:9733-9743.

Silva M, Benito A, Sanz C, Prosper F, Ekhterae D, Nunez G, Fernandez-Luna JL (1999) Erythropoietin can induce the expression of bcl-x(L) through Stat5 in erythropoietin-dependent progenitor cell lines. J Biol Chem 274:22165-22169.

Sinor AD, Greenberg DA (2000) Erythropoietin protects cultured cortical neurons, but not astroglia, from hypoxia and AMPA toxicity. Neurosci Lett 290:213-215.

Siren AL, Knerlich F, Poser W, Gleiter CH, Bruck W, Ehrenreich H (2001a) Erythropoietin and erythropoietin receptor in human ischemic/hypoxic brain. Acta Neuropathol (Berl) 101:271-276.

Siren AL, Fratelli M, Brines M, Goemans C, Casagrande S, Lewczuk P, Keenan S, Gleiter C, Pasquali C, Capobianco A, Mennini T, Heumann R, Cerami A, Ehrenreich H, Ghezzi P (2001b) Erythropoietin prevents neuronal apoptosis after cerebral ischemia and metabolic stress. Proc Natl Acad Sci U S A 98:4044-4049.

Slater P, McConnell SE, D'Souza SW, Barson AJ (1993) Postnatal changes in N-methyl-D-aspartate receptor binding and stimulation by glutamate and glycine of [3H]-MK-801 binding in human temporal cortex. Br J Pharmacol 108:1143-1149.

Solaroglu I, Solaroglu A, Kaptanoglu E, Dede S, Haberal A, Beskonakli E, Kilinc K (2003) Erythropoietin prevents ischemia-reperfusion from inducing oxidative damage in fetal rat brain. Childs Nerv Syst 19:19-22.

Springborg JB, Ma X, Rochat P, Knudsen GM, Amtorp O, Paulson OB, Juhler M, Olsen NV (2002) A single subcutaneous bolus of erythropoietin normalizes cerebral blood flow autoregulation after subarachnoid haemorrhage in rats. Br J Pharmacol 135:823-829.

Stave U (1965) Age-dependent changes of metabolism. II. Influences of hypoxia on tissue enzyme patterns of newborn and adult rabbits. Biol Neonat 8:114-130.

Stewart WB, Ment LR, Schwartz M (1997) Chronic postnatal hypoxia increases the numbers of cortical neurons. Brain Res 760:17-21.

Subramaniam S, McGonigle P (1994) Regional profile of developmental changes in the sensitivity of the N-methyl-D-aspartate receptor to polyamines. J Neurochem 62:1408-1415.

Sugawa M, Sakurai Y, Ishikawa-Ieda Y, Suzuki H, Asou H (2002) Effects of erythropoietin on glial cell development; oligodendrocyte maturation and astrocyte proliferation. Neurosci Res 44:391-403.

Tabira T, Konishi Y, Gallyas F, Jr. (1995) Neurotrophic effect of hematopoietic cytokines on cholinergic and other neurons in vitro. Int J Dev Neurosci 13:241-252.

Teramo KA, Widness JA, Clemons GK, Voutilainen P, McKinlay S, Schwartz R (1987) Amniotic fluid erythropoietin correlates with umbilical plasma erythropoietin in normal and abnormal pregnancy. Obstet Gynecol 69:710-716.

Tsuji M, Saul JP, du Plessis A, Eichenwald E, Sobh J, Crocker R, Volpe JJ (2000) Cerebral intravascular oxygenation correlates with mean arterial pressure in critically ill premature infants. Pediatrics 106:625-632.

Vaccarino FM, Ment LR (2004) Injury and repair in developing brain. Arch Dis Child Fetal Neonatal Ed 89:F190-192.

Vairano M, Dello Russo C, Pozzoli G, Battaglia A, Scambia G, Tringali G, Aloe-Spiriti MA, Preziosi P, Navarra P (2002) Erythropoietin exerts anti-apoptotic effects on rat microglial cells in vitro. Eur J Neurosci 16:584-592.

Vallance P, Benjamin N, Collier J (1988) Erythropoietin, haemoglobin, and hypertensive crises. Lancet 1:1107.

Vannucci RC (1990) Experimental biology of cerebral hypoxia-ischemia: relation to perinatal brain damage. Pediatr Res 27:317-326.

Vannucci RC (1993) Experimental models of perinatal hypoxic-ischemic brain damage. APMIS Suppl 40:89-95.

Vannucci RC, Perlman JM (1997) Interventions for perinatal hypoxic-ischemic encephalopathy. Pediatrics 100:1004-1014.

Vannucci RC, Vannucci SJ (1997) A model of perinatal hypoxic-ischemic brain damage. Ann N Y Acad Sci 835:234-249.

Villa P, Bigini P, Mennini T, Agnello D, Laragione T, Cagnotto A, Viviani B, Marinovich M, Cerami A, Coleman TR, Brines M, Ghezzi P (2003) Erythropoietin selectively attenuates cytokine production and inflammation in cerebral ischemia by targeting neuronal apoptosis. J Exp Med 198:971-975.

Volpe JJ (2001a) Neurobiology of periventricular leukomalacia in the premature infant. Pediatr Res 50:553-562.

Volpe JJ (2001b) Perinatal brain injury: From pathogenesis to neuroprotection. Ment Retard Dev Disabil Res Rev 7:56-64.

Volpe JJ (2001c) Neurology of the Newborn, 4th Edition: W.B. Saunders.

Wang CH, Liang CL, Huang LT, Liu JK, Hung PH, Sun A, Hung KS (2004) Single intravenous injection of naked plasmid DNA encoding erythropoietin provides neuroprotection in hypoxia-ischemia rats. Biochem Biophys Res Commun 314:1064-1071.

Wiessner C, Allegrini PR, Ekatodramis D, Jewell UR, Stallmach T, Gassmann M (2001) Increased cerebral infarct volumes in polyglobulic mice overexpressing erythropoietin. J Cereb Blood Flow Metab 21:857-864.

Wolf RF, Gilmore LS, Friese P, Downs T, Burstein SA, Dale GL (1997a) Erythropoietin potentiates thrombus development in a canine arterio-venous shunt model. Thromb Haemost 77:1020-1024.

Wolf RF, Peng J, Friese P, Gilmore LS, Burstein SA, Dale GL (1997b) Erythropoietin administration increases production and reactivity of platelets in dogs. Thromb Haemost 78:1505-1509.

Wong KC, Li PK, Lui SF, Nicholls MG, Lai KN (1990) The adverse effects of recombinant human erythropoietin therapy. Adverse Drug React Acute Poisoning Rev 9:183-206.

Wrighton NC, Farrell FX, Chang R, Kashyap AK, Barbone FP, Mulcahy LS, Johnson DL, Barrett RW, Jolliffe LK, Dower WJ (1996) Small peptides as potent mimetics of the protein hormone erythropoietin. Science 273:458-464.

Wu H, Liu X, Jaenisch R, Lodish HF (1995) Generation of committed erythroid BFU-E and CFU-E progenitors does not require erythropoietin or the erythropoietin receptor. Cell 83:59-67.

Wu H, Lee SH, Gao J, Liu X, Iruela-Arispe ML (1999) Inactivation of erythropoietin leads to defects in cardiac morphogenesis. Development 126:3597-3605.

Yan XX, Najbauer J, Woo CC, Dashtipour K, Ribak CE, Leon M (2001) Expression of active caspase-3 in mitotic and postmitotic cells of the rat forebrain. J Comp Neurol 433:4-22.

Yasuda Y, Nagao M, Okano M, Masuda S, Sasaki R, Konishi H, al e (1993) Localization of erythropoietin and erythropoietin-receptor in postimplantation mouse embryos. Development, Growth and Differentiation 35:711-722.

Yonezawa M, Back SA, Gan X, Rosenberg PA, Volpe JJ (1996) Cystine deprivation induces oligodendroglial death: rescue by free radical scavengers and by a diffusible glial factor. J Neurochem 67:566-573.

Yoon BH, Romero R, Kim CJ, Koo JN, Choe G, Syn HC, Chi JG (1997) High expression of tumor necrosis factor-alpha and interleukin-6 in periventricular leukomalacia. Am J Obstet Gynecol 177:406-411.

Yu X, Lin CS, Costantini F, Noguchi CT (2001) The human erythropoietin receptor gene rescues erythropoiesis and developmental defects in the erythropoietin receptor null mouse. Blood 98:475-477.

Yu X, Shacka JJ, Eells JB, Suarez-Quian C, Przygodzki RM, Beleslin-Cokic B, Lin CS, Nikodem VM, Hempstead B, Flanders KC, Costantini F, Noguchi CT (2002) Erythropoietin receptor signalling is required for normal brain development. Development 129:505-516.

Zipursky A (2002) The risk of hematopoietic growth factor therapy in newborn infants. Pediatr Res 51:549.

Chapter 6

ERYTHROPOIETIN FOR THE TREATMENT OF ACUTE ISCHEMIC STROKE: PRECLINICAL RATIONALE

Michael J. Renzi, Linda K. Jolliffe, Francis X. Farrell and Kenneth J. Rhodes
Johnson & Johnson Pharmaceutical Research and Development, L.L.C., New jersey, USA

Abstract: Ischemic stroke is a result of the disruption of blood flow in the cerebral circulation that leads to neuronal dysfunction and cell death. Approved treatment of ischemic stroke is limited to thrombolysis using TPA. The future of stroke therapy is focused on preventing cell death and enhancing endogenous recovery mechanisms to improve outcome. Erythropoietin has received considerable attention for its ability to promote neuronal survival following an insult. EPO mediated neuroprotection has been demonstrated in several *in vitro* and *in vivo* models relevant to ischemic stroke. While the neuroprotective function of EPO is thought to be primarily through the inhibition of apoptosis, more recent work has revealed additional functions for EPO that may contribute to its beneficial effects including the modulation of inflammation, preventing vascular dysfunction and enhancing angiogenesis and neurogenesis. This chapter will provide an overview of the pre-clinical research that supports the potential use of EPO in treating stroke. The effects of EPO on the mechanisms that contribute to damage following ischemic stroke will be discussed as well as more recent work describing the effect of EPO treatment on recovery of function.

Key words: Ischemia, apoptosis, inflammation, and recovery.

1. INTRODUCTION: ISCHEMIC STROKE

Stroke is caused by a disruption of the flow of blood to the brain that most often results in impaired neurological function. The type and severity of impairment depends on the size of the lesion and the area of the brain that is affected. Strokes are typically classified as hemorrhagic and ischemic.

Hemorrhagic stroke, the result of bleeding from a blood vessel, can occur in the brain itself or in the meningeal layers that surround the brain. Ischemic stroke refers to the physical occlusion of a blood vessel by an obstruction such as a thrombus or an embolus. Both hemorrhagic and ischemic stroke result in a decrease in blood flow, depriving the area of the brain supplied by the occluded or bleeding vessel of oxygen, glucose and other essential nutrients. The pathophysiology of stroke is complex and the following discussion will focus on mechanisms of cell injury related to ischemic stroke, acknowledging that some of these mechanisms are also relevant to hemorrhagic stroke.

The loss of blood flow to the brain initiates a cascade of events that, if unresolved, will lead to cell death in the effected area. Because many areas of the brain receive a portion of their blood supply via highly collateralized vascular networks, the occlusion of a single vessel may produce a variable decrease in blood flow in downstream tissue resulting in an equally variable severity of insult. The area where the reduction in blood flow is the greatest is called the *core* of the stroke and cell death in this region is for the most part rapid, irreversible and necrotic in nature. The region surrounding the core is termed the *penumbra*. In this area the reduction in blood flow is not as severe and as a result, cell death is delayed and occurs through mechanisms that are more heterogeneous than those taking place in the core. Without intervention, most cells within the penumbra will succumb over time and the area of tissue loss will expand outward from the core of the lesion. Restoration of blood flow is vital to preserving tissue, however, reperfusion itself initiates a series of mechanisms that can also lead to cell death, a process called reperfusion injury. Therefore the successful treatment of ischemic stroke must not only restore blood flow to the affected areas, but must target the mechanisms that have been initiated by the initial ischemia and the reperfusion injury. Considerable attention has been focused on defining the ischemic penumbra and elucidating mechanisms that contribute to the expanding area of cell death, thereby defining targets for therapeutic intervention.

The initial cell death following ischemic stroke is quite rapid, occurring in the minutes to hours following the initial insult. The inability of the neuron to maintain the necessary energy levels in an environment of reduced oxygen and glucose is the major factor contributing to early cell death. Energy loss leads to the neuron's inability to maintain membrane potential and thereby maintain ionic gradients (Hansen and Nedergaard, 1988). As these gradients break down, ions, including Ca^{2+}, enter the cell and initiate a series of events that result in release of excitatory neurotransmitters such as glutamic acid. The subsequent over activation of ionotropic glutamate receptors leads to the uncontrolled influx of cations into neurons, resulting in the death of those neurons, a process called excitotoxicity (Aarts and Tymianski, 2004; Choi and Rothman, 1990). Increased levels of intracellular Ca^{2+} also activates proteases that degrade the cytoskeletal matrix as well as other enzymes that lead to the liberation of toxic mediators including reactive oxygen species (Choi, 1990; Sugawara and Chan, 2003).

The expansion of cell death into the penumbra occurs over the subsequent hours to days and during this time apoptosis and inflammation play a significant role (for reviews see (Chamorro and Planas, 2004; Choi, 1998; Love, 2003). Apoptosis, or programmed cell death, is an energy dependant cell death pathway that requires the synthesis and activation of specific genes and proteins. The amount of cell death that is contributed by apoptotic mechanisms is unclear. Apoptosis is thought to occur mainly in the penumbra where the insult is less severe. This observation is consistent with results in animal models that show that apoptotic cell death is more prevalent in models of stroke that result in a larger penumbral zone. (Chen et al., 1997; Ferrer and Planas, 2003; Hakim, 1998). In human stroke patients, the contribution of apoptosis to delayed cell death is even less clear, with post-mortem analysis revealing few cells that exhibit classic signs of apoptosis (e.g., condensed or fragmented nuclei; membrane blebbing, etc.) (Love et al., 2000). Inflammation following stroke is the result of both the activation of astrocytes and microglia within the CNS as well as the infiltration of neutrophils, macrophages and monocytes from the systemic circulation (Stoll et al., 2002). Each of these cell types produces reactive oxygen species and releases cytokines and other cytotoxic inflammatory mediators (Ishikawa et al., 2004). Some of these mediators also injure vascular endothelial cells, resulting in the activation of matrix proteases and contributing to the disruption of the blood brain barrier (for review see (del Zoppo and Hallenbeck, 2000). This barrier disruption may lead to the further influx of inflammatory cells, edema, and an increased risk for hemorrhagic transformation.

In addition to the processes that contribute to cell death and tissue loss, endogenous processes of remodeling and repair also occur following stroke and can result in some degree of preservation or restoration of function. Examples include the proliferation, migration and differentiation of endogenous neural progenitor cells (Felling and Levison, 2003; Li et al., 2002; Zhang et al., 2004). In addition, increased expression of markers of neurite outgrowth (Kawamata et al., 1997; Stroemer et al., 1998) and synapse formation (Chen et al., 2003; Stroemer et al., 1998) have also been reported in animals subject to focal ischemic stroke. Vascular remodeling in the ischemic penumbra can also occur, suggesting that the blood supply can also be repaired in the effected area (Zhang et al., 2003). The cellular processes involved in reconstruction and remodeling are attractive targets for new stroke therapies owing at least in part to the prolonged time frame over which they occur, thus providing a potentially large time window for therapeutic intervention.

Although the various mechanisms for cell death provide a wealth of potential drug targets and opportunities for therapeutic intervention, significant challenges remain. One challenge is posed by the observation that the various cell death pathways, once initiated, proceed in parallel. Therefore an inhibitor acting at a single check point in a single cell death pathway may attenuate cell death but may not provide sufficient protection to provide a meaningful effect on stroke outcome. A second challenge is posed by the observation that some of the pathways for cell death require a single trigger and then are rapidly amplified via kinase or protease activation, resulting in recruitment and activation of several different downstream effector enzymes. As a result, inhibition of far upstream "initiator" enzymes may be required to effect meaningful protection. Due to the complexity of the pathological process that contributes to neuronal death following stroke it may be more beneficial to target multiple mechanisms with a single therapeutic molecule or combinations of molecules. This strategy is becoming increasingly attractive, as single drug therapies have consistently failed in stroke trials. An ideal therapy would be a single molecule that has the ability to affect several cell death or cell survival signals at once, and also have the ability to trigger a broad spectrum of responses leading to proliferation of neuronal and vascular stem cells. Erythropoietin may represent this type of molecule. The remainder of this chapter will focus on the pre-clinical evidence that supports erythropoietin as a novel therapy to treat ischemic stroke.

2. ERYTHROPOIETIN AS A NEUROPROTECTANT

Erythropoietin (EPO) is a 34 kDa glycoprotein that is primarily responsible for controlling the production of red blood cells (Koury and Bondurant, 1990). EPO binds to its receptor located on the surface of erythroid precursor cells and stimulates their proliferation, differentiation and survival during hematopoiesis. As has been described in previous chapters, the receptor for EPO is expressed on the surface of cells in the central nervous system (CNS) including on neurons, astrocytes, microglia and oligodendrocytes (Nagai et al., 2001).

EPO is produced by the kidney in response to low circulating levels of oxygen in the blood. Hypoxia can also induce the production of EPO in the CNS, with astrocytes producing EPO in response to low oxygen levels (Masuda et al., 1994). Levels of EPO receptor also increase on neurons and on astrocytes in response to various types of injury, including hypoxia and ischemia (Siren et al., 2001b). The endogenous EPO/EPOR system resident in the CNS therefore, is responsive to hypoxic and other insults, and several reports indicate that interfering with this response results in increased neuronal death (Sakanaka et al., 1998). Further support for the hypothesis that EPO plays a role in neuronal survival following hypoxic injury comes from the observation that EPO is responsible, at least in part, for the protective effects of ischemic tolerance. Ischemic tolerance, or ischemic preconditioning, describes a condition whereby a small ischemic event can lead to a reduction in the damage that occurs following a subsequent, more severe event (Kitigawa et al., 1990). EPO has been shown to mimic the effect of a preconditioning stimulus in an in vitro model of tolerance (Ruscher et al., 2002). Infusion of soluble EPO receptor to neutralize EPO signaling attenuates the beneficial effects of ischemic preconditioning (Prass et al., 2003) providing further evidence for the role of EPO in this response and for a more general protective effect of EPO in the CNS.

The first example of exogenously administered EPO reducing the damage in an animal model of stroke was reported in 1998 (Sadamoto et al., 1998). In this study, continuous intracerebroventricular (i.c.v.) administration of EPO reduced the cortical infarction, loss of thalamic neurons and behavioral deficits in a rodent model of permanent middle cerebral artery occlusion (MCAO). A similar beneficial effect of EPO was observed in a permanent MCAO model in the mouse (Bernaudin et al., 1999). Importantly, in 2000 Brines et al., reported that peripheral (intraperitoneal) administration of EPO (5000 IU/kg) reduced the damage following permanent MCAO in the rat (Brines et al., 2000), demonstrating that systemically administered EPO effectively protects the brain following injury. In this same study, labeled EPO was administered intraperitoneally and subsequently detected in the CNS, showing for the first time that EPO administered systemically could cross the blood brain barrier and gain access to the brain following injury. Subsequent work has confirmed that EPO can also cross an intact BBB (Juul et al., 2004); other reports suggests that EPO can be actively transported across the BBB through EPO receptors located on the surface of cerebrovascular endothelial cells (Brines et al., 2000; Eid et al., 2004). The Brines report also began to define a therapeutic time window for EPO in focal ischemic stroke, demonstrating that EPO treatment given as late as 6 hr after the ischemic insult still resulted in a statistically significant reduction in infarct volume.

Elucidation of the cellular mechanisms underlying the protective effects of EPO in neurons remains an active area of research. Several studies have reported on the effects of EPO in the CNS, and this work points to a number of mechanisms through which EPO may provide cytoprotection. Sites of action include direct effects on neurons, modulation of inflammatory responses by affecting glial activation and preserving the function of cerebrovascular endothelial cells and stimulation of the survival and differentiation of endogenous neuronal progenitor cells. A better understanding of the mechanisms through which EPO provides cytoprotection will help in designing dosing strategies to maximize the beneficial effects of EPO in the CNS.

2.1 Direct Effects on Neurons

EPO can affect neurons directly as evidenced by the expression of the EPO receptor on the plasma membrane of neurons, the stimulation of EPOR phosphorylation in response to EPO binding and the activation of downstream signaling events in response to EPOR activation. Previous reports have described the neuronal responses to EPOR activation, including up-regulation of anti-apoptotic proteins including BCL2, activation of JAK/STAT phosphorylation and stabilization of HIF-1. These responses may explain, in part, the protective effect of EPO on neurons.

As described earlier, several processes contribute to the damage following stroke and excitotoxicity is one of the early mechanisms. EPO has been demonstrated to block neuronal death in several models of excitotoxicity, including toxicity associated with exposure to high concentrations of glutamate and NMDA (Morishita et al., 1997; Sakanaka et al., 1998). EPO does not appear to block excitotoxic cell death by inhibiting the over-activation of the NMDA receptor directly but instead targets downstream processes that are the result of excessive Ca^{2+} influx (Chong et al., 2003a; Digicaylioglu and Lipton, 2001). The activity of EPO in these models typically requires pretreatment for at least 8 hours, indicating that RNA and protein synthesis is necessary for protection against glutamate challenge. EPO has also been shown to block excitotoxic neuronal death in a model of chemical ischemia (Kawakami et al., 2001). In hippocampal slices, the protective effects of EPO are mediated through presynaptic inhibition of glutamate release induced by ischemia. It is not known how EPO regulates glutamate release, but the effect is rapid and does not require pretreatment, indicating that that no new protein synthesis was required. Attenuating the release of glutamate could slow down or halt the propagation of excitotoxicity. EPO therefore has the potential to inhibit excitotoxicity by two distinct mechanisms, one that inhibits glutamate release pre-synaptically, and one that attenuates the neighboring cells' response to glutamate.

EPO has also been shown to affect Ca^{2+} homeostasis in neurons. While EPO does not appear to directly modulate Ca^{2+} influx through the NMDA receptor in models of excitotoxicity, EPO treatment has been shown to promote Ca^{2+} entry into neurons, possibly through activity involving T-type Ca^{2+} channels (Assandri et al., 1999; Masuda et al., 1993). In fact, increases in intracellular Ca^{2+} in response to EPO treatment has been proposed as the likely mechanism behind EPO mediated protection of PC-12 cells following NGF withdrawal (Koshimura et al., 1999). How EPO mediated increases in intracellular Ca^{2+} promotes cell survival is not known.

Perhaps the most interesting of EPO's effects on neurons in the context of ischemic stroke and neuroprotection, is the ability to promote cell survival by inhibiting apoptosis. During hematopoiesis, EPO regulates apoptosis to control the number of erythroid progenitor cells that mature to become red blood cells. By blocking apoptosis, EPO promotes cell survival resulting ultimately in an increase in erythrocyte number. Recent work has demonstrated that the anti-apoptotic activity of EPO extends to other cell types including cardiac myocytes (Tramontano et al., 2003) endothelial cells (Carlini et al., 1999) and neurons. EPO can effectively block apoptosis in neuronal-like cell lines as well as in primary cultures of cortical, hippocampal and cerebellar granule neurons exposed to a variety of stimuli relevant to stroke pathology. These stimuli include chemical hypoxia (Wen et al., 2002), free radical damage (Chong et al., 2003b), oxygen glucose deprivation (Ruscher et al., 2002), NMDA (Digicaylioglu and Lipton, 2001), serum withdrawal, kainic acid treatment (Siren et al., 2001a) and hypoxia (Lewczuk et al., 2000). Much of the evidence supporting an anti-apoptotic effect of EPO in neurons comes from studies of cultures enriched for neurons, indicating that the anti-apoptotic involve direct activation of neuronal receptors. In animal models of focal ischemic stroke, EPO has also been shown to decrease the number of neurons undergoing apoptosis following an ischemic event (Siren et al., 2001a), suggesting that the observations made using *in vitro* assays extend to the *in vivo* setting. The mechanisms underlying the anti-apoptotic effects of EPO in neurons are not completely clear but likely involve activation of several signaling pathways. Many of the signaling pathways and genes that are activated in response to EPO exposure in neurons are involved in apoptosis. Examples include phosphorylation and activation of AKT through PI3 Kinase signaling, (Chong et al., 2003a), activation and nuclear translocation of $NF_{k}\beta$ (Digicaylioglu and Lipton, 2001), activation of MAPK and PKB (Siren et al., 2001a) and expression and phosphorylation of members of the Bcl-2 family of proteins including Bad (Chong et al., 2003b), bak and Bcl_{XL} (Renzi et al., 2002; Wen et al., 2002). The signaling pathways that lead to neuroprotective effects are covered in more detail in chapter 3. Direct effects on neurons are only one possible site of action through which EPO may exert its neuroprotective function. There are other cell types involved in the pathological and restorative processes, as described previously, these cells are responsive to EPO and may be critical to the overall neuroprotective and pro-regenerative properties of this cytokine.

2.2 Effects on Astrocytes and Microglia

Inflammation is recognized as a process that contributes significantly to the damage that occurs following ischemic stroke. Previous work has shown that EPO can reduce the inflammation that occurs in models of spinal cord injury (Gorio et al., 2002) and in models of experimental autoimmune encephalitis (Agnello et al., 2002; Brines et al., 2000). Does EPO have similar anti-inflammatory effects following ischemic stroke?

The activation of astrocytes and microglia in the CNS following ischemic injury has been well documented. The ultimate consequences of the activation of these cell types remains unclear. The activation of astrocytes and microglia can have both positive and negative consequences for neurons following ischemia. Both astrocytes and microglia secrete cytokines that are pro-inflammatory and neurotoxic; however, both cell types also secrete neuroprotective growth factors. Both cells types function to maintain ionic homeostasis in the brain and also help maintain the integrity of the neurovascular unit. Therefore, any discussion of modulating glial cell function following ischemic stroke must consider both the deleterious and beneficial functions of these cell types, and also their effects on the cerebral vasculature.

The EPO receptor is present on the surface of both astrocytes and microglia and the expression appears to increase, at least in the case of astrocytes, following injury (Siren et al., 2001b). A number of investigators have studied the effect of EPO on astrocytes and microglia, evaluating production and release of growth factors, proliferation and cell survival. EPO has been shown to elevate glutathione peroxidase levels in cultured astrocytes, suggesting that it may function to modulate injury caused by free radicals. There is, however, no evidence that EPO can directly modulate the activation and subsequent pro-inflammatory functions of these two cell types. Villa et al., 2003, observed that EPO was unable to inhibit LPS-stimulated TNFα production by cultured rat glial cells in culture (Villa et al., 2003). They did find, however, that TNFα was released and glial cells were activated in neuron/glia co-cultures subjected to neurotoxin exposure. EPO treatment in these cultures led to cell survival and an inhibition of glial activation. These results are consistent with those published by (Chong et al., 2002), who observed microglia activation following exposure to conditioned media obtained from neurons exposed to NO. Treatment with EPO prior to NO exposure prevented neuronal cell death and also prevented the activation of microglia by the conditioned media. *In vivo*, some studies reported that EPO has anti-inflammatory effects as evidenced by a decreased inflammatory response in models of experimental autoimmune encephalitis (EAE) and spinal cord injury (Agnello et al., 2002; Gorio et al., 2002). Moreover, following ischemic stroke, Villa demonstrated that EPO treatment resulted in a significant decrease in inflammation as measured by a decrease in astrocyte and microglia activation, by reduced leukocyte infiltration into the infarcted area, and by decreased production of the pro-inflammatory cytokines IL-6, TNF and MCP-1 (Villa et al., 2003). Together, these studies suggest that the ability of EPO to modulate inflammation, particularly microglia activation, is indirect, and that the effects on inflammatory responses may be a consequence of cytoprotection rather than a direct anti-inflammatory effect per se.

2.3 Effects on Endothelial cells and the cerebral vasculature

Another potential site of the beneficial effects of EPO following ischemic stroke is in the cerebral vasculature. Vascular endothelial cells express the EPOR, and EPO has been shown to have a direct effect on these cells. EPO treatment blocks LPS induced apoptosis of endothelial cells in culture (Carlini et al., 1999) and there is evidence that EPO can improve endothelial cell function in chronic renal failure patients (Kuriyama et al., 1996). In brain, micro vascular endothelial cells are particularly sensitive to ischemia (for review see (del Zoppo and Hallenbeck, 2000). Briefly, ischemia can cause endothelial cell dysfunction that results in a loss of tight junctions and compromise of the BBB, the loss of extracellular matrix, increased leukocyte adhesion and the release of secreted factors that control vascular tone. These events can lead to edema, hemorrhage and inflammation, factors that contribute to the expansion of injury and tissue loss following ischemic stroke. Following ischemic stroke, these vasculature-protective effects of EPO would be of added benefit.

EPO has also been shown to have a protective effect on endothelial cells of the brain microvasculature. Endothelial cells grown in culture and exposed to anoxia or free radical damage are protected from apoptosis by EPO treatment (Chong et al., 2002). The mechanism by which EPO elicits this anti-apoptotic response appears to involve pathways that are similar to the anti-apoptotic pathways in neurons, and include the activation of akt-1, increased Bcl_{XL} expression and downstream effects on mitochondrial function and caspase activity. EPO has also been shown to preserve vascular function in animal models of sub-arachnoid hemorrhage including, where it may decrease vasospasm and preserve cerebral blood flow (for review see (Grasso et al., 2002).

2.4 Effects on Stem Cells – Neurogenesis and angiogenesis

The potential role of EPO in activating endogenous repair processes following stroke has provided intriguing new insight into the possible mechanisms that may underlie the beneficial effects of EPO following ischemic stroke. For example, Michael Chopp and colleagues have reported that EPO treatment following embolic stroke stimulates the proliferation and differentiation of endogenous neuronal progenitor cells and also stimulates angiogenesis, leading to formation of new capillaries within the ischemic penumbra (Wang et al., 2004). These responses are mirrored by increased release of VEGF and other neuroprotective, pro-angiogenic growth factors. These cellular responses to EPO also correspond to improved behavioral outcome, as rats treated with EPO following embolic MCAO stroke show improved performance in neurologic function and in performance of complex behavioral tasks that assess sensorimotor coordination and integration.

3. CONSIDERATIONS ON DOSING REGIMEN

The data described above support the potential utility of EPO for treating ischemic stroke. Determining the ideal dosing regimen will require considering how EPO is eliciting its effects.

As summarized above EPO likely acts though several different mechanisms to improve the outcome following ischemic stroke. While considerable attention has focused on the its anti-apoptotic properties EPO has also been shown to function in a number of other relevant processes including endothelial cell function and mechanisms of remodeling and reconstruction including angiogenesis and neurogenesis. The timing at which these event occur range from hours following the ischemic insult to days and potentially weeks after, suggesting that an optimum regimen might involve early administration of EPO followed by administration at regular intervals following the injury.

The amount of EPO to give is no less of a mystery. Considerations include the amount of EPO that is able to get into the brain, the integrity of the blood brain barrier in the hours to days following stroke and the observation by a number of labs that EPO activity can be lost when the concentration is too high (inverted U-shaped dose response curve). EPO has been used successfully to improve outcome in a pilot study of ischemic stroke (Ehrenreich et al., 2002) however only a single dose level of EPO was investigated. Dosing ranging as well as PK experiments should be carried out to determine if this is the optimum regimen. Also critical is the effect of dosing regimen on hematopoiesis. Ideally, one would like to deliver EPO in a manner that provides neuroprotection and stimulates repair processes without raising hemoglobin or increasing hematocrit.

4. SUMMARY

In summary, EPO is an exciting molecule for CNS applications and may represent a transformational new therapy for the treatment of ischemic stroke among other types of CNS injury. Future work will need to carefully explore dose levels and establish relationships between dosing and drug effect. These additional data, coupled with the in vitro and in vivo animal model data described above will enable the stroke community to transfer EPO from the laboratory into clinical use.

REFERENCES

Aarts, M. M., and Tymianski, M. (2004). Molecular mechanisms underlying specificity of excitotoxic signaling in neurons. Curr Mol Med *4*, 137-147.

Agnello, D., Bigini, P., Villa, P., Mennini, T., Cerami, A., Brines, M. L., and Ghezzi, P. (2002). Erythropoietin exerts an anti-inflammatory effect on the CNS in a model of experimental autoimmune encephalomyelitis. Brain Res *952*, 128-134.

Assandri, R., Egger, M., Gassmann, M., Niggli, E., Bauer, C., Forster, I., and Gorlach, A. (1999). Erythropoietin modulates intracellular calcium in a human neuroblastoma cell line. J Physiol *516 (Pt 2)*, 343-352.

Bernaudin, M., Marti, H. H., Roussel, S., Divoux, D., Nouvelot, A., MacKenzie, E. T., and Petit, E. (1999). A potential role for erythropoietin in focal permanent cerebral ischemia in mice. J Cereb Blood Flow Metab *19*, 643-651.

Brines, M. L., Ghezzi, P., Keenan, S., Agnello, D., de Lanerolle, N. C., Cerami, C., Itri, L. M., and Cerami, A. (2000). Erythropoietin crosses the blood-brain barrier to protect against experimental brain injury. Proc Natl Acad Sci U S A *97*, 10526-10531.

Carlini, R. G., Alonzo, E. J., Dominguez, J., Blanca, I., Weisinger, J. R., Rothstein, M., and Bellorin-Font, E. (1999). Effect of recombinant human erythropoietin on endothelial cell apoptosis. Kidney Int *55*, 546-553.

Chamorro, A., and Planas, A. M. (2004). Inflammation-mediated damage as a potential therapeutic target in acute ischemic stroke. Ernst Schering Res Found Workshop, 185-204.

Chen, J., Jin, K., Chen, M., Pei, W., Kawaguchi, K., Greenberg, D. A., and Simon, R. P. (1997). Early detection of DNA strand breaks in the brain after transient focal ischemia:

implications for the role of DNA damage in apoptosis and neuronal cell death. J Neurochem *69*, 232-245.

Chen, J., Zhang, Z. G., Li, Y., Wang, Y., Wang, L., Jiang, H., Zhang, C., Lu, M., Katakowski, M., Feldkamp, C. S., and Chopp, M. (2003). Statins induce angiogenesis, neurogenesis, and synaptogenesis after stroke. Ann Neurol *53*, 743-751.

Choi, D. (1998). Antagonizing excitotoxicity: a therapeutic strategy for stroke? Mt Sinai J Med *65*, 133-138.

Choi, D. W. (1990). Cerebral hypoxia: some new approaches and unanswered questions. J Neurosci *10*, 2493-2501.

Choi, D. W., and Rothman, S. M. (1990). The role of glutamate neurotoxicity in hypoxic-ischemic neuronal death. Annu Rev Neurosci *13*, 171-182.

Chong, Z. Z., Kang, J. Q., and Maiese, K. (2002). Erythropoietin is a novel vascular protectant through activation of Akt1 and mitochondrial modulation of cysteine proteases. Circulation *106*, 2973-2979.

Chong, Z. Z., Kang, J. Q., and Maiese, K. (2003a). Erythropoietin fosters both intrinsic and extrinsic neuronal protection through modulation of microglia, Akt1, Bad, and caspase-mediated pathways. Br J Pharmacol *138*, 1107-1118.

Chong, Z. Z., Lin, S. H., Kang, J. Q., and Maiese, K. (2003b). Erythropoietin prevents early and late neuronal demise through modulation of Akt1 and induction of caspase 1, 3, and 8. J Neurosci Res *71*, 659-669.

del Zoppo, G. J., and Hallenbeck, J. M. (2000). Advances in the vascular pathophysiology of ischemic stroke. Thromb Res *98*, 73-81.

Digicaylioglu, M., and Lipton, S. A. (2001). Erythropoietin-mediated neuroprotection involves cross-talk between Jak2 and NF-kappaB signalling cascades. Nature *412*, 641-647.

Ehrenreich, H., Hasselblatt, M., Dembowski, C., Cepek, L., Lewczuk, P., Stiefel, M., Rustenbeck, H. H., Breiter, N., Jacob, S., Knerlich, F., *et al.* (2002). Erythropoietin therapy for acute stroke is both safe and beneficial. Mol Med *8*, 495-505.

Eid, T., Brines, M. L., Cerami, A., Spencer, D. D., Kim, J. H., Schweitzer, J. S., Ottersen, O. P., and de Lanerolle, N. C. (2004). Increased expression of erythropoietin receptor on blood vessels in the human epileptogenic hippocampus with sclerosis. J Neuropathol Exp Neurol *63*, 73-83.

Felling, R. J., and Levison, S. W. (2003). Enhanced neurogenesis following stroke. J Neurosci Res *73*, 277-283.

Ferrer, I., and Planas, A. M. (2003). Signaling of cell death and cell survival following focal cerebral ischemia: life and death struggle in the penumbra. J Neuropathol Exp Neurol *62*, 329-339.

Gorio, A., Gokmen, N., Erbayraktar, S., Yilmaz, O., Madaschi, L., Cichetti, C., Di Giulio, A. M., Vardar, E., Cerami, A., and Brines, M. (2002). Recombinant human erythropoietin counteracts secondary injury and markedly enhances neurological recovery from experimental spinal cord trauma. Proc Natl Acad Sci U S A *99*, 9450-9455.

Grasso, G., Buemi, M., Alafaci, C., Sfacteria, A., Passalacqua, M., Sturiale, A., Calapai, G., De Vico, G., Piedimonte, G., Salpietro, F. M., and Tomasello, F. (2002). Beneficial effects of systemic administration of recombinant human erythropoietin in rabbits subjected to subarachnoid hemorrhage. Proc Natl Acad Sci U S A *99*, 5627-5631.

Hakim, A. M. (1998). Ischemic penumbra: the therapeutic window. Neurology *51*, S44-46.

Hansen, A. J., and Nedergaard, M. (1988). Brain ion homeostasis in cerebral ischemia. Neurochem Pathol *9*, 195-209.

Ishikawa, M., Zhang, J. H., Nanda, A., and Granger, D. N. (2004). Inflammatory responses to ischemia and reperfusion in the cerebral microcirculation. Front Biosci *9*, 1339-1347.

Juul, S. E., McPherson, R. J., Farrell, F. X., Jolliffe, L., Ness, D. J., and Gleason, C. A. (2004). Erytropoietin concentrations in cerebrospinal fluid of nonhuman primates and fetal sheep following high-dose recombinant erythropoietin. Biol Neonate *85*, 138-144.

Kawakami, M., Sekiguchi, M., Sato, K., Kozaki, S., and Takahashi, M. (2001). Erythropoietin receptor-mediated inhibition of exocytotic glutamate release confers neuroprotection during chemical ischemia. J Biol Chem *276*, 39469-39475.

Kawamata, T., Dietrich, W. D., Schallert, T., Gotts, J. E., Cocke, R. R., Benowitz, L. I., and Finklestein, S. P. (1997). Intracisternal basic fibroblast growth factor enhances functional recovery and up-regulates the expression of a molecular marker of neuronal sprouting following focal cerebral infarction. Proc Natl Acad Sci U S A *94*, 8179-8184.

Koshimura, K., Murakami, Y., Sohmiya, M., Tanaka, J., and Kato, Y. (1999). Effects of erythropoietin on neuronal activity. J Neurochem *72*, 2565-2572.

Koury, M. J., and Bondurant, M. C. (1990). Erythropoietin retards DNA breakdown and prevents programmed death in erythroid progenitor cells. Science *248*, 378-381.

Kuriyama, S., Hopp, L., Yoshida, H., Hikita, M., Tomonari, H., Hashimoto, T., and Sakai, O. (1996). Evidence for amelioration of endothelial cell dysfunction by erythropoietin therapy in predialysis patients. Am J Hypertens *9*, 426-431.

Lewczuk, P., Hasselblatt, M., Kamrowski-Kruck, H., Heyer, A., Unzicker, C., Siren, A. L., and Ehrenreich, H. (2000). Survival of hippocampal neurons in culture upon hypoxia: effect of erythropoietin. Neuroreport *11*, 3485-3488.

Li, Y., Chen, J., and Chopp, M. (2002). Cell proliferation and differentiation from ependymal, subependymal and choroid plexus cells in response to stroke in rats. J Neurol Sci *193*, 137-146.

Love, S. (2003). Apoptosis and brain ischaemia. Prog Neuropsychopharmacol Biol Psychiatry *27*, 267-282.

Love, S., Barber, R., and Wilcock, G. K. (2000). Neuronal death in brain infarcts in man. Neuropathol Appl Neurobiol *26*, 55-66.

Masuda, S., Nagao, M., Takahata, K., Konishi, Y., Gallyas, F., Jr., Tabira, T., and Sasaki, R. (1993). Functional erythropoietin receptor of the cells with neural characteristics. Comparison with receptor properties of erythroid cells. J Biol Chem *268*, 11208-11216.

Masuda, S., Okano, M., Yamagishi, K., Nagao, M., Ueda, M., and Sasaki, R. (1994). A novel site of erythropoietin production. Oxygen-dependent production in cultured rat astrocytes. J Biol Chem *269*, 19488-19493.

Morishita, E., Masuda, S., Nagao, M., Yasuda, Y., and Sasaki, R. (1997). Erythropoietin receptor is expressed in rat hippocampal and cerebral cortical neurons, and erythropoietin prevents in vitro glutamate-induced neuronal death. Neuroscience *76*, 105-116.

Nagai, A., Nakagawa, E., Choi, H. B., Hatori, K., Kobayashi, S., and Kim, S. U. (2001). Erythropoietin and erythropoietin receptors in human CNS neurons, astrocytes, microglia, and oligodendrocytes grown in culture. J Neuropathol Exp Neurol *60*, 386-392.

Prass, K., Scharff, A., Ruscher, K., Lowl, D., Muselmann, C., Victorov, I., Kapinya, K., Dirnagl, U., and Meisel, A. (2003). Hypoxia-induced stroke tolerance in the mouse is mediated by erythropoietin. Stroke *34*, 1981-1986.

Renzi, M. J., Farrell, F. X., Bittner, A., Galindo, J. E., Morton, M., Trinh, H., and Jolliffe, L. K. (2002). Erythropoietin induces changes in gene expression in PC-12 cells. Brain Res Mol Brain Res *104*, 86-95.

Ruscher, K., Freyer, D., Karsch, M., Isaev, N., Megow, D., Sawitzki, B., Priller, J., Dirnagl, U., and Meisel, A. (2002). Erythropoietin is a paracrine mediator of ischemic tolerance in the brain: evidence from an in vitro model. J Neurosci *22*, 10291-10301.

Sadamoto, Y., Igase, K., Sakanaka, M., Sato, K., Otsuka, H., Sakaki, S., Masuda, S., and Sasaki, R. (1998). Erythropoietin prevents place navigation disability and cortical infarction in rats with permanent occlusion of the middle cerebral artery. Biochem Biophys Res Commun *253*, 26-32.

Sakanaka, M., Wen, T. C., Matsuda, S., Masuda, S., Morishita, E., Nagao, M., and Sasaki, R. (1998). In vivo evidence that erythropoietin protects neurons from ischemic damage. Proc Natl Acad Sci U S A *95*, 4635-4640.

Siren, A. L., Fratelli, M., Brines, M., Goemans, C., Casagrande, S., Lewczuk, P., Keenan, S., Gleiter, C., Pasquali, C., Capobianco, A., *et al.* (2001a). Erythropoietin prevents neuronal apoptosis after cerebral ischemia and metabolic stress. Proc Natl Acad Sci U S A *98*, 4044-4049.

Siren, A. L., Knerlich, F., Poser, W., Gleiter, C. H., Bruck, W., and Ehrenreich, H. (2001b). Erythropoietin and erythropoietin receptor in human ischemic/hypoxic brain. Acta Neuropathol (Berl) *101*, 271-276.

Stoll, G., Jander, S., and Schroeter, M. (2002). Detrimental and beneficial effects of injury-induced inflammation and cytokine expression in the nervous system. Adv Exp Med Biol *513*, 87-113.

Stroemer, R. P., Kent, T. A., and Hulsebosch, C. E. (1998). Enhanced neocortical neural sprouting, synaptogenesis, and behavioral recovery with D-amphetamine therapy after neocortical infarction in rats. Stroke *29*, 2381-2393; discussion 2393-2385.

Sugawara, T., and Chan, P. H. (2003). Reactive oxygen radicals and pathogenesis of neuronal death after cerebral ischemia. Antioxid Redox Signal *5*, 597-607.

Tramontano, A. F., Muniyappa, R., Black, A. D., Blendea, M. C., Cohen, I., Deng, L., Sowers, J. R., Cutaia, M. V., and El-Sherif, N. (2003). Erythropoietin protects cardiac myocytes from hypoxia-induced apoptosis through an Akt-dependent pathway. Biochem Biophys Res Commun *308*, 990-994.

Villa, P., Bigini, P., Mennini, T., Agnello, D., Laragione, T., Cagnotto, A., Viviani, B., Marinovich, M., Cerami, A., Coleman, T. R., *et al.* (2003). Erythropoietin selectively attenuates cytokine production and inflammation in cerebral ischemia by targeting neuronal apoptosis. J Exp Med *198*, 971-975.

Wang, L., Zhang, Z., Wang, Y., Zhang, R., and Chopp, M. (2004). Treatment of stroke with erythropoietin enhances neurogenesis and angiogenesis and improves neurological function in rats. Stroke *35*, 1732-1737.

Wen, T. C., Sadamoto, Y., Tanaka, J., Zhu, P. X., Nakata, K., Ma, Y. J., Hata, R., and Sakanaka, M. (2002). Erythropoietin protects neurons against chemical hypoxia and cerebral ischemic injury by up-regulating Bcl-xL expression. J Neurosci Res *67*, 795-803.

Zhang, B., Wang, R. Z., Yao, Y., Liu, Z. H., Lian, Z. G., Zou, Y. J., and Wei, Y. K. (2004). Proliferation and differentiation of neural stem cells in adult rats after cerebral infarction. Chin Med Sci J *19*, 73-77.

Zhang, R., Wang, L., Zhang, L., Chen, J., Zhu, Z., Zhang, Z., and Chopp, M. (2003). Nitric oxide enhances angiogenesis via the synthesis of vascular endothelial growth factor and cGMP after stroke in the rat. Circ Res *92*, 308-313.

Chapter 7

ERYTHROPOIETIN NEUROPROTECTION IN THE RETINA

Gundula Rohde, Mathias Bähr and Jochen H. Weishaupt
Center for Neurological Medicine, University of Göttingen, Robert-Koch-Str. 40, 37075 Göttingen, Germany

Abstract: The cytokine erythropoietin (EPO) regulates production of red blood cells in an oxygen-dependent manner by inhibition of apoptosis of erythrocyte precursors in the bone marrow. EPO and EPO receptor protein are also expressed by CNS neurons and glial cells. EPO exerts neuroprotective effects in various experimental in vitro and in vivo models of neural injury, such as mechanical trauma, neuroinflammation, and cerebral and retinal ischemia. In this chapter, we will focus on the effects of EPO on lesioned retinal neurons, and discuss its potential value for treatment of retinal diseases. We will also summarize recent data regarding EPO signal transduction that underlies neuroprotection in retinal neuronal cell death models, including optic nerve transection or retinal ischemia.

Key words: Apoptosis, neuronal cell death, retinal ganglion cells, neurite outgrowth, retinal ischemia

1. INTRODUCTION

Recent studies show that cellular effects of EPO upon binding to its receptor on erythrocyte precursor, are mediated by activation of the Akt, MAPK p42/44, and JAK-STAT pathways (Kashii et al., 2000; Sakamoto et al., 2000; Uddin et al., 2000; Ratajczak et al., 2001), resulting in increased expression levels of anti-apoptotic proteins such as Bcl-2 or Bcl-Xl (Socolovsky et al., 2001; Dolznig et al., 2002). In vitro studies indicate a significant overlap between EPO signalling in erythropoid progenitors and neuronal cells, as both the Akt pathway and, less consistently between different paradigms, the Ras/Raf/ERK1/2 kinase cascade, are involved in EPO signal transduction. Moreover, in an in vitro model of neuronal excitotoxicity, a crosstalk between Janus-tyrosine kinase-2 (JAK2) kinase and NF-κB was found to be necessary for survival-promoting EPO effects (Digicaylioglu and Lipton, 2001). However, our understanding of the neuroprotective EPO signal transduction, especially in vivo, is still limited.

In the following chapter, we summarize our understanding of the neuroprotective properties of EPO using retinal lesion models, including optic nerve transection, retinal ischemia and light-induced retinal degeneration. We outline advantages of retinal paradigms for neuronal cell death research, which include the clear retinal tissue layering that facilitates identification of different cell types. Moreover, different application routes for pharmacological agents, including direct non-systemic access by intravitreal injections, and the fact that retinal ganglion cells represent a subpopulation of CNS neurons, allow elegant studies on apoptosis of CNS neurons in the optic nerve transection paradigm. Respective in vitro models, such as immunopurified retinal ganglion cells or retinal explants, expand the experimental usefulness of retinal cell death paradigms. We delineate important findings concerning survival-promoting effects and signal transduction of EPO in these models, and discuss possible implications for new therapeutic approaches in neurological and ophthalmological diseases.

2. EPO RECEPTOR EXPRESSION IN THE RETINA

The erythropoietin receptor (EPOR) is expressed on the surface of erythroid precursor cells. It belongs to the cytokine receptor superfamily, for which substantial information about signal transduction cascades exist. It shares several structural features with other family members (Bazan, 1990). The domain structure of this receptor family consists of an extracellular domain, which binds EPO, a single transmembrane domain and an intracellular domain responsible for signaling. Upon EPO binding, EPOR forms homodimers, which allows its autophosphorylation by the receptor-associated JAK2. Subsequently, several different signaling pathways are activated, including PI-3-kinase/Akt, ERK1/2, STAT5-dependent gene transcription or JAK2-dependent NF-κB activation.

In recent years, EPOR was found to be expressed in mouse neural and brain tissue during embryonal development (Liu et al., 1994; Liu et al., 1997), as well as in neuronal and glial cells of adult hippocampus, cerebral cortex and cerebellum of rodents, monkeys and humans (Digicaylioglu et al., 1995; Marti et al., 1996; Morishita et al., 1997). Moreover, EPOR was found to be expressed in the developing human central nervous system (Li et al.,

1996; Juul et al., 1998a) as well as in cultured neurons and astrocytes (Konishi et al., 1993; Masuda et al., 1994; Marti et al., 1996; Juul et al., 1998b; Bernaudin et al., 1999; Bernaudin et al., 2000; Bocker-Meffert et al., 2002).

Figure 7-1. Expression of erythropoietin receptor (EPOR) protein in the rat retinal ganglion cell layer. (A), EPOR immunostaining of the retinal ganglion cell layer using a Cy3-labeled secondary antibody shows EPOR expression on neuronal somata and dendrites (arrowheads). (B), retrograde staining of retinal ganglion cells by injection of fluorogold into the superior colliculus, demonstrating EPOR expression predominantly in retinal ganglion cells (RGCs).

Substantial changes in EPO and EPOR expression have been demonstrated following various types of CNS injuries, they are discussed elsewhere in the book. Several studies demonstrated the expression of EPOR protein and mRNA in the mammalian retina, including human tissue specimen. Western blot experiments showed EPOR protein expression in mouse and rat retina at different postnatal stages up to 2 months of age. Immunohistochemical experiments revealed EPOR immunoreactivity in photoreceptor cells (rod inner segment) as well as outer plexiform layer, inner plexiform layer and to a lesser extend in the inner nuclear layer of rats (Sprague Dawley and Fisher strain) and mice (BALB/c strain) (Bocker-Meffert et al., 2002; Grimm et al., 2002; Junk et al., 2002; Grimm et al., 2004; Weishaupt et al., 2004). Minimal (Junk et al., 2002) to robust (Bocker-

Meffert et al., 2002; Weishaupt et al., 2004) EPOR expression was reported in retinal ganglion cells of control animals (Fig. 7.1). Further EPOR protein upregulation was noted 60 min. after retinal ischemia in retinal ganglion cells, retinal astrocytes and amacrine neurons (Junk et al., 2002). This is in accordance with an increase in neuronal and glial EPOR expression which was shown to be induced by ischemia or hypoxia in the brain, suggesting a possible role for endogenous EPO in the retina's and brain's response to injury (Sakanaka et al., 1998; Ehrenreich et al., 2002). In contrast, optic nerve transection, i.e. axotomy of rat retinal ganglion cells, did not result in a change in EPOR expression in retinal ganglion cells as shown by Western blotting and immunohistochemistry (Weishaupt et al., 2004). Therefore, it appears that the upregulation of EPOR expression following a lesion may not be an obligate observation, but may depend on the type or acuteness of the insult, or even be a specific response to hypoxia.

3. NEUROPROTECTIVE EFFECTS OF EPO ON RETINAL CELLS

As already mentioned above, EPO has been proven to protect neurons and exert anti-apoptotic activity in several settings of neuronal cell death, e.g. in vitro oxygen-glucose deprivation (Grimm et al., 2002), cerebral ischemia (Ihle et al., 1993; Ehrenreich et al., 2002; Chong et al., 2003a), spinal cord ischemia (Grimm et al., 2002), or even human stroke (Uddin et al., 2000). In the first study on biological effects of EPO on retinal ganglion cells, Böcker-Meffert and colleagues showed that EPO (and VEGF) enhances retinal ganglion cell neurite outgrowth in postnatal rat retinal explants; this effect could be blocked by the calcium channel blocker ESX or neutralising antibodies (Bocker-Meffert et al., 2002). Since then, in accordance with the widespread expression of EPOR protein in the retina, several studies demonstrate that EPO has a neuroprotective effect on various retinal cell types.

In 2002, Grimm and colleagues demonstrated that recombinant human EPO (rhEPO) protects mouse photoreceptors from light-induced programmed cell death and reduces caspase activation (Grimm et al., 2002). Hypoxic preconditioning induced HIF-1α and EPO expression, prevented photoreceptor degeneration and preserved retinal function, strengthening the hypothesis that HIF-1α-induced upregulation of EPO expression contributes to the protective effect of ischemic preconditioning in the CNS. In fact, systemically applied rhEPO, given before or after the apoptosis-inducing light stimulus, could mimic the protective effect of hypoxic preconditioning and demonstrated that EPO has the capability to cross the blood-retina barrier. Most recently, it was also shown that transgenic overexpression (Grimm et al., 2004) of human EPO protected photoreceptor cells from light-induced degeneration. In contrast, EPO, both systemically applied or using EPO overexpressing mice, was not effective against photoreceptor degeneration in two genetic mouse models of human inherited retinitis pigmentosa (Grimm et al., 2004), suggesting different apoptotic pathways in light-induced or genetic photoreceptor degeneration.

In parallel, Junk and colleagues published a study on the effects of EPO in the context of retinal ischemia, a serious and common ophtalmological problem (Junk et al., 2002). In this mouse model of transient global retinal ischemia, induced by increasing intraocular pressure to 120 mm Hg for 45 or 60 min., widespread neuronal degeneration, positive TUNEL staining and retinal thinning was observed. The retinal layers most affected in this paradigm were inner retinal layers, specifically the inner nuclear layer. Systemic administration of rhEPO, even when given after the insult, inhibited neuronal apoptosis induced by retinal ischemia throughout all retinal layers, including photoreceptors and the retinal ganglion cell layer. Moreover, EPO preserved retinal function as shown by electroretinogram (ERG) measurements. In accordance with prior data showing that intraventricular infusion of soluble EPO receptor potentiated neuronal damage, intravitreal injection of EPO neutralising EPOR protein resulted in a further decreased ERG amplitude after retinal ischemia, again supporting a possible role of EPO in endogenous neuroprotective mechanisms.

Recently, axotomy of rat retinal ganglion cells (RGCs) was used to study EPO neuroprotection as well as signal transduction in vivo (Weishaupt et al., 2004). RGCs, regarding their developmental history as well as their response to injury and regeneration inhibiting factors are part of the CNS. Following transection of the optic nerve, i.e. axotomy of RGCs, about 90% of the RGCs die within 14 days. During a short procedure that includes supraorbital surgical access, the optic nerve is transected about 5 mm behind the eyeball (Fig. 7-2). RGCs are specifically labelled by placing gel foam soaked in a solution of a retrograde fluorescent tracer, usually fluorogold or DiI, at the distal nerve stump. After 14 days, retinae are flat mounted and surviving RGCs can be counted using a fluorescent microscope, expressed as cells/mm2 (Fig. 7-4) (Garcia-Valenzuela et al., 1994). Substances that may influence neuronal degeneration can be either injected directly into the vitreous body with direct access to the retinal ganglion cells, or can be applied systemically (Kermer et al., 2001). This in vivo paradigm does not suffer from the high complexity and less clear-cut and reproducible neuronal loss of other models for neuronal degeneration in the CNS, while all aspects of neuronal degeneration can be studied on a cellular and molecular level. Retrograde death of axotomized RGCs is associated with characteristic features of apoptosis. Besides classical morphological signs of apoptotic cell death (Bien et al., 1999), they include activation of caspase-3 and -9 (Kermer et al., 1999b; Kermer et al., 2000b), antagonistic regulation of Bax and Bcl-2 expression (Isenmann et al., 1997), downregulation of apoptosis-inhibiting kinase activity (Kermer et al., 2000a; Klöcker et al., 2000) and protective effects of caspase inhibitors (Kermer et al., 1998; Kermer et al., 1999a; Kermer et al., 2000a) as well as neurotrophins (Klöcker et al., 1998; Klöcker et al., 1999; Kermer et al., 2000a; Klöcker et al., 2000).

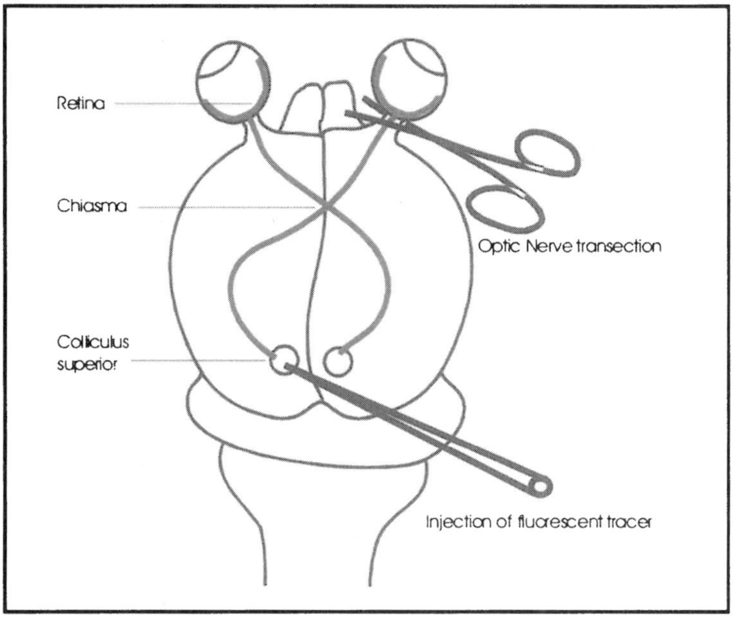

Figure 7-2. Schematic drawing of the rat brain, demonstrating optic nerve transection in rats and retrograde fluorogold labelling of retinal ganglion cells by injection of the retrograde fluorescent tracer into the superior colliculus

Repeated intraocular EPO injections almost doubled the number of surviving RGCs 14 days after optic nerve transection in rats, and significantly reduced the number of RGC immunopositive for activated caspase-3 (Weishaupt et al., 2004). Similar to other neuronal injury models, different concentrations of EPO revealed a bell-shaped dose-response curve. However, although protective effects were abrogated at highest doses tested, toxic effects, i.e. RGC counts below control numbers, were not observed. Thus, profound EPO neuroprotection against axotomy-induced death of CNS neurons could be shown using a subacute to chronic in vivo cell death paradigm with exclusively apoptotic features, whereas to date most in vivo data describing EPO neuroprotection were obtained from acute hypoxia/ischemia paradigms with predominantly necrotic types of cell death. Moreover, this model recapitulates crucial steps of glaucoma pathophysiology, because axonal lesions due to increased ocular tension are thought to induce retrograde RGC death and subsequent vision loss (Kikuchi et al., 2000; Pease et al., 2000; Levin et al., 2001). However, care must be taken when applying the results of the optic nerve transection model to human glaucoma, as this model, in contrast to other animal models described for this disease, lacks other aspects of glaucoma such as elevation of intraocular pressure (Levin et al., 2001).

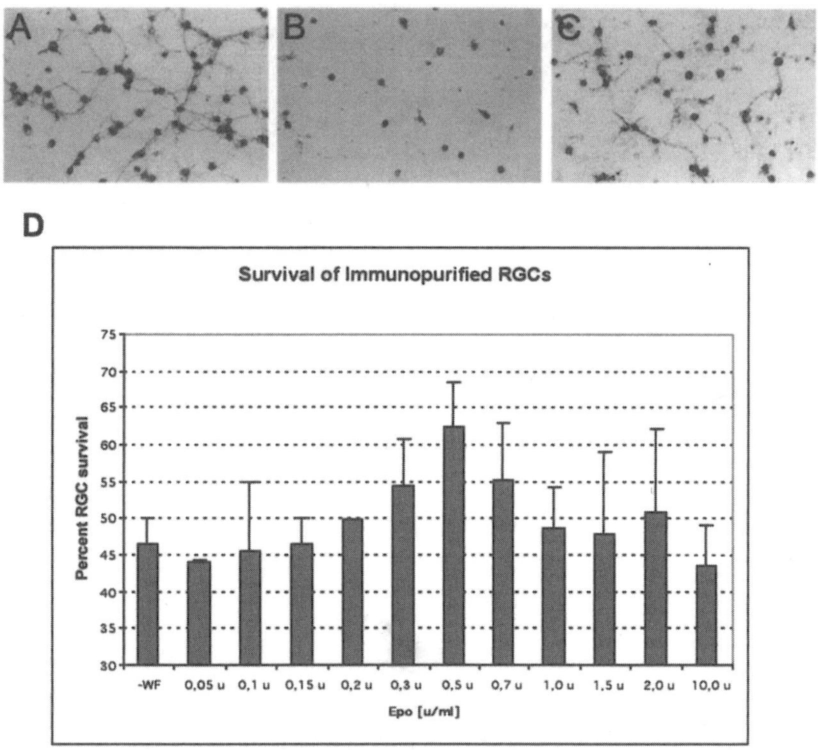

Figure 7-3. Erythropoietin promotes survival and neurite outgrowth of immunopurified rat retinal ganglion cell after neurotrophic factor deprivation. (A), RGCs from 6-8-days-old rat pups were immunopurified and cultured under full neurotrophic support for 24 hours. (B), vehicle-treated RGCs after two additional days of neurotrophic factor deprivation. (C), neurotrophic factor deprivation and treatment with EPO (0.6 U/ml) for 2 days. Panel below: Quantification and dose-response-curve (percent surviving RGCs compared to full neurotrophic support).

EPO receptors are also expressed by glial cells (Koury and Bondurant, 1990; Masuda et al., 1994; Sadamoto et al., 1998; Bernaudin et al., 1999).

Thus, contribution of secondary protective effects due to involvement of non-neuronal cell types cannot be excluded in vivo. However, an *in vitro* model, corresponding to the in vivo RGC axotomy paradigm, showed that EPO exerts direct neuroprotection on retinal ganglion cells. In a two-step immunopurification protocol (Barres et al., 1988), RGCs from P7 rats are purified to near homogeneity and cultured in serum-free medium, supplemented with forskolin, human BDNF, CNTF and insulin as neurotrophic factors. Apoptotic RGC death was induced by neurotrophin withdrawal after 1 day in culture. Two days after neurotrophic factor withdrawal, the number of surviving RGCs was determined by MTT assay. In this purely neuronal cell culture model, EPO enhanced the survival of immunopurified RGCs significantly. Again, as also found in the in vivo optic nerve transection model and various other neuronal cell culture models, different EPO doses resulted in a bell-shaped dose-response curve, with the maximal RGC survival at an EPO concentration of approximately 0.5 U/ml (Fig. 7-3). Moreover, in accordance with the observations by Bocker-Meffert and colleagues who found that EPO increased neurite extension in retinal explants (Bocker-Meffert et al., 2002), EPO treated immunopurified retinal ganglion cells had a clearly enhanced neurite number and length (Fig. 7-3).

4. EPO SIGNAL TRANSDUCTION IN THE RETINA

Signal transduction of the EPO/EPOR system was initially characterized in erythrocyte precursor cells. In these non-neuronal cells, binding of EPO triggers EPO receptor dimerization resulting in JAK2 activation. Further downstream, three different signal transduction pathways are predominantly activated: the PI3K/Akt pathway, the RAS/RAF/ERK (ERK1/2) pathway as well as transcriptional activation of STAT5 (Signal Transducer and Activator of Transcription 5), which, upon phosphorylation by JAK2, translocates to the nucleus and initiates transcription of anti-apoptotic genes such as Bcl-Xl (Yoshimura and Misawa, 1998). Overall, similar signal transduction pathways for EPO survival signalling in neurons seem to be shared. The PI3K/Akt and the ERK1/2 pathways have been shown to transduce EPO effects in cultured cortical or hippocampal neurons (Siren et al., 2001; Bouscary et al., 2003; Chong et al., 2003b). Akt signalling was found to be more important than ERK1/2 activation, in agreement with the observation that the latter kinase pathway promotes proliferation rather than survival in the myeloid system. In general, the importance of the MAPK p42/44 (ERK1/2) kinase pathway for cell survival has remained ambiguous. In neurons, this kinase cascade seems to be more relevant in the context of neurite extension and differentiation (Kim et al., 1997; Fukunaga and

Miyamoto, 1998; Leppa et al., 1998; Robinson et al., 1998; Aletsee et al., 2001; Barnabe-Heider and Miller, 2003).

Recently, as a new aspect in EPO signal transduction, Digicaylioglu and Lipton identified JAK2 mediated NFκB activation as being a necessary event for protection of cerebrocortical neurons against NO- and NMDA-mediated cell death (Digicaylioglu and Lipton, 2001).

In general, in vivo data concerning neuronal EPO signalling is scarce, at least partially due to technical limitations, making it difficult to apply kinase inhibitors to brain tissue, which would allow to test the functional relevance of kinase pathways found to be activated by EPO treatment. Due to the easier accessibility of the retina by intravitreal injections of neuroprotective molecules, kinase inhibitors, other pharmacological compounds or even TAT fusion proteins and viral vectors, the retinal system turned out to be an elegant model to investigate effects and signal transduction of neuroprotective molecules (Weishaupt et al., 2003a; Weishaupt et al., 2003b).

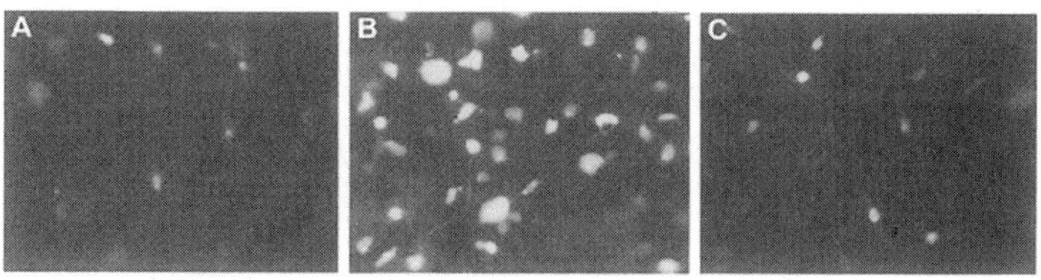

Figure 7-4. Wortmannin blocks EPO neuroprotection in the retina. (A), Retinal whole mount with retrogradely fluorogold-labeled RGCs 14 days after axotomy under vehicle treatment. (B), retinal whole-mount 14 days after ON transection under treatment with 2 U EPO/eye, injected on days 0, 4, 7 and 10 after axotomy. (C), retinal whole mount after co-treatment with EPO and WM 14 days after ON transection.

While Grimm and colleagues observed reduced retinal ERK1/2 phosphorylation after preconditioning ischemia (Grimm et al., 2002), the phosphorylation of Akt or expression patterns of the possible downstream players Bcl-Xl, XIAP and NFκB remained unchanged. However, exogenously applied EPO injected into the vitreous body induced phosphorylation of Akt in the optic nerve transection study (Weishaupt et al., 2004) at concentrations shown to protect RGCs from axotomy-induced apoptosis. Moreover, under combined treatment with EPO and wortmannin, an inhibitor of the PI-3-K/Akt pathway, the neuroprotective effect of EPO was completely abolished (Fig. 7-4). No change in protein expression or phosphorylation of other signal transduction kinases that are involved in EPO-induced signalling and apoptosis regulation in other paradigms, were observed. Neither p38 nor JNK displayed any difference regarding the level of their phosphorylated form or total protein, and, similar to the findings by Grimm and colleagues in preconditioning ischemia, retinal levels of bcl-2, Bcl-Xl or XIAP were not affected by EPO treatment.

Thus, recent evidence supports the view that the PI-3-kinase/Akt pathway is the main, if not sole, mediator of EPO neuroprotection, at least in the rodent retina.

5. CONCLUSION

In summary, EPO neuroprotection has been investigated in 4 different retinal in vivo models, which substantially extended the understanding of EPO as a neuroprotectant. Its therapeutic effects have been demonstrated against retrograde degeneration of axotomized rat retinal ganglion cells, retinal ischemia in rats, light induced photoreceptor degeneration in mice as well as in mouse models for retinitis pigmentosa. Data from the retinal ganglion cell axotomy model have been further substantiated by in vitro experiments using immunopurified rat retinal ganglion cells.

Concerning basic insights into neurobiology of ophthalmologic and neurological diseases, at least two different apoptotic pathways are possible in photoreceptor cells, one of which can be inhibited by EPO. However, the second apoptotic pathway proceeds independent of transgenic EPO expression or application of exogenous EPO (Grimm et al., 2004). While, EPO is effective in light-induced photoreceptor degeneration, EPO has no effect in mouse models of inherited retinitis pigmentosa.

In general, the retinal ganglion cell axotomy model is a highly reproducible and precise model for neuronal cell death in the CNS, and the above-described studies using rodent retina further underline the therapeutic potential of EPO to prevent neuronal cell death. In vitro experiments using retinal explants or immunopurified retinal ganglion cells suggest that EPO has not only anti-apoptotic effects, but may also enhance axonal regeneration (Fig. 7-3; Bocker-Meffert et al., 2002).The bell-shaped EPO dose response curve previously seen in several neuronal cell culture models could be confirmed in vivo using the optic nerve transection paradigm, which might have relevance for the design of upcoming clinical trials. The therapeutic effects of EPO in clinical studies might be further improved by an optimized dosing regime in the future.

Of course, even more concrete conclusions can be drawn from the above experiments for the treatment of retinal diseases, as both retinal ischemia and spontaneous or inherited photoreceptor degeneration are common clinical problems in ophthalmology. The fact that EPO is effective given systemically, demonstrates that, at least in rodents, the blood-retina barrier is permissive for this cytokine allowing the possibility of a systemic mode of application in clinical practice. In addition, the optic nerve transection model recapitulates important pathophysiological steps relevant for RGC death and vision loss due to glaucoma, because axonal lesions due to increased ocular tension are known to induce apoptotic RGC death and subsequent vision loss (Kikuchi et al., 2000; Pease et al., 2000; Levin, 2001).

REFERENCES

Aletsee C, Beros A, Mullen L, Palacios S, Pak K, Dazert S, Ryan AF (2001) Ras/MEK but not p38 signaling mediates NT-3-induced neurite extension from spiral ganglion neurons. J Assoc Res Otolaryngol 2:377-387.

Barnabe-Heider F, Miller FD (2003) Endogenously produced neurotrophins regulate survival and differentiation of cortical progenitors via distinct signaling pathways. J Neurosci 23:5149-5160.

Barres BA, Silverstein BE, Corey DP, Chun LL (1988) Immunological, morphological, and electrophysiological variation among retinal ganglion cells purified by panning. Neuron 1:791-803.

Bazan JF (1990) Haemopoietic receptors and helical cytokines. Immunol Today 11:350-354.

Bernaudin M, Marti HH, Roussel S, Divoux D, Nouvelot A, MacKenzie ET, Petit E (1999) A potential role for erythropoietin in focal permanent cerebral ischemia in mice. J Cereb Blood Flow Metab 19:643-651.

Bernaudin M, Bellail A, Marti HH, Yvon A, Vivien D, Duchatelle I, Mackenzie ET, Petit E (2000) Neurons and astrocytes express EPO mRNA: oxygen-sensing mechanisms that involve the redox-state of the brain. Glia 30:271-278.

Bien A, Seidenbecher CI, Bockers TM, Sabel BA, Kreutz MR (1999) Apoptotic versus necrotic characteristics of retinal ganglion cell death after partial optic nerve injury. J Neurotrauma 16:153-163.

Bocker-Meffert S, Rosenstiel P, Rohl C, Warneke N, Held-Feindt J, Sievers J, Lucius R (2002) Erythropoietin and VEGF promote neural outgrowth from retinal explants in postnatal rats. Invest Ophthalmol Vis Sci 43:2021-2026.

Bouscary D, Pene F, Claessens YE, Muller O, Chretien S, Fontenay-Roupie M, Gisselbrecht S, Mayeux P, Lacombe C (2003) Critical role for PI 3-kinase in the control of erythropoietin-induced erythroid progenitor proliferation. Blood 101:3436-3443.

Chong ZZ, Kang JQ, Maiese K (2003a) Apaf-1, Bcl-xL, cytochrome c, and caspase-9 form the critical elements for cerebral vascular protection by erythropoietin. J Cereb Blood Flow Metab 23:320-330.

Chong ZZ, Lin SH, Kang JQ, Maiese K (2003b) Erythropoietin prevents early and late neuronal demise through modulation of Akt1 and induction of caspase 1, 3, and 8. J Neurosci Res 71:659-669.

Digicaylioglu M, Lipton SA (2001) Erythropoietin-mediated neuroprotection involves cross-talk between Jak2 and NF-kappaB signalling cascades. Nature 412:641-647.

Digicaylioglu M, Bichet S, Marti HH, Wenger RH, Rivas LA, Bauer C, Gassmann M (1995) Localization of specific erythropoietin binding sites in defined areas of the mouse brain. Proc Natl Acad Sci U S A 92:3717-3720.

Dolznig H, Habermann B, Stangl K, Deiner EM, Moriggl R, Beug H, Mullner EW (2002) Apoptosis protection by the Epo target Bcl-X(L) allows factor-independent differentiation of primary erythroblasts. Curr Biol 12:1076-1085.

Ehrenreich H, Hasselblatt M, Dembowski C, Cepek L, Lewczuk P, Stiefel M, Rustenbeck HH, Breiter N, Jacob S, Knerlich F, Bohn M, Poser W, Ruther E, Kochen M, Gefeller O, Gleiter C, Wessel TC, De Ryck M, Itri L, Prange H, Cerami A, Brines M, Siren AL (2002) Erythropoietin therapy for acute stroke is both safe and beneficial. Mol Med 8:495-505.

Fukunaga K, Miyamoto E (1998) Role of MAP kinase in neurons. Mol Neurobiol 16:79-95.

Garcia-Valenzuela E, Gorczyca W, Darzynkiewicz Z, Sharma SC (1994) Apoptosis in adult retinal ganglion cells after axotomy. J Neurobiol 25:431-438.

Grimm C, Wenzel A, Groszer M, Mayser H, Seeliger M, Samardzija M, Bauer C, Gassmann M, Reme CE (2002) HIF-1-induced erythropoietin in the hypoxic retina protects against light-induced retinal degeneration. Nat Med 8:718-724.

Grimm C, Wenzel A, Stanescu D, Samardzija M, Hotop S, Groszer M, Naash M, Gassmann M, Reme C (2004) Constitutive overexpression of human erythropoietin protects the mouse retina against induced but not inherited retinal degeneration. J Neurosci 24:5651-5658.

Ihle JN, Quelle FW, Miura O (1993) Signal transduction through the receptor for erythropoietin. Semin Immunol 5:375-389.

Isenmann S, Wahl C, Krajewski S, Reed JC, Bähr M (1997) Up-regulation of Bax protein in degenerating retinal ganglion cells precedes apoptotic cell death after optic nerve lesion in the rat. Eur J Neurosci 9:1763-1772.

Junk AK, Mammis A, Savitz SI, Singh M, Roth S, Malhotra S, Rosenbaum PS, Cerami A, Brines M, Rosenbaum DM (2002) Erythropoietin administration protects retinal neurons from acute ischemia-reperfusion injury. Proc Natl Acad Sci U S A 99:10659-10664.

Juul SE, Yachnis AT, Christensen RD (1998a) Tissue distribution of erythropoietin and erythropoietin receptor in the developing human fetus. Early Hum Dev 52:235-249.

Juul SE, Anderson DK, Li Y, Christensen RD (1998b) Erythropoietin and erythropoietin receptor in the developing human central nervous system. Pediatr Res 43:40-49.

Kashii Y, Uchida M, Kirito K, Tanaka M, Nishijima K, Toshima M, Ando T, Koizumi K, Endoh T, Sawada K, Momoi M, Miura Y, Ozawa K, Komatsu N (2000) A member of Forkhead family transcription factor, FKHRL1, is one of the downstream molecules of phosphatidylinositol 3-kinase-Akt activation pathway in erythropoietin signal transduction. Blood 96:941-949.

Kermer P, Klöcker N, Bähr M (1999a) Long-term effect of inhibition of ced 3-like caspases on the survival of axotomized retinal ganglion cells in vivo. Exp Neurol 158:202-205.

Kermer P, Klöcker N, Labes M, Bähr M (1998) Inhibition of CPP32-like proteases rescues axotomized retinal ganglion cells from secondary cell death in vivo. J Neurosci 18:4656-4662.

Kermer P, Klöcker N, Labes M, Bähr M (2000a) Insulin-like growth factor-I protects axotomized rat retinal ganglion cells from secondary death via PI3-K-dependent Akt phosphorylation and inhibition of caspase-3 In vivo. J Neurosci 20:2-8.

Kermer P, Klöcker N, Labes M, Thomsen S, Srinivasan A, Bähr M (1999b) Activation of caspase-3 in axotomized rat retinal ganglion cells in vivo. FEBS Lett 453:361-364.

Kermer P, Klöcker N, Weishaupt JH, Bähr M (2001) Transection of the optic nerve in rats: studying neuronal death and survival in vivo. Brain Res Brain Res Protoc 7:255-60.

Kermer P, Ankerhold R, Klöcker N, Krajewski S, Reed JC, Bähr M (2000b) Caspase-9: involvement in secondary death of axotomized rat retinal ganglion cells in vivo. Brain Res Mol Brain Res 85:144-150.

Kikuchi M, Tenneti L, Lipton SA (2000) Role of p38 mitogen-activated protein kinase in axotomy-induced apoptosis of rat retinal ganglion cells. J Neurosci 20:5037-5044.

Kim B, Leventhal PS, Saltiel AR, Feldman EL (1997) Insulin-like growth factor-I-mediated neurite outgrowth in vitro requires mitogen-activated protein kinase activation. J Biol Chem 272:21268-21273.

Klöcker N, Cellerino A, Bähr M (1998) Free radical scavenging and inhibition of nitric oxide synthase potentiates the neurotrophic effects of brain-derived neurotrophic factor on axotomized retinal ganglion cells In vivo. J Neurosci 18:1038-1046.

Klöcker N, Kermer P, Gleichmann M, Weller M, Bähr M (1999) Both the neuronal and inducible isoforms contribute to upregulation of retinal nitric oxide synthase activity by brain-derived neurotrophic factor. J Neurosci 19:8517-8527.

Klöcker N, Kermer P, Weishaupt JH, Labes M, Ankerhold R, Bähr M (2000) Brain-derived neurotrophic factor-mediated neuroprotection of adult rat retinal ganglion cells in vivo does not exclusively depend on phosphatidyl-inositol-3'-kinase/protein kinase B signaling. J Neurosci 20:6962-6967.

Konishi Y, Chui DH, Hirose H, Kunishita T, Tabira T (1993) Trophic effect of erythropoietin and other hematopoietic factors on central cholinergic neurons in vitro and in vivo. Brain Res 609:29-35.

Koury MJ, Bondurant MC (1990) Erythropoietin retards DNA breakdown and prevents programmed death in erythroid progenitor cells. Science 248:378-381.

Kuperstein F, Yavin E (2002) ERK activation and nuclear translocation in amyloid-beta peptide- and iron-stressed neuronal cell cultures. Eur J Neurosci 16:44-54.

Leppa S, Saffrich R, Ansorge W, Bohmann D (1998) Differential regulation of c-Jun by ERK and JNK during PC12 cell differentiation. Embo J 17:4404-4413.

Levin LA (2001) Animal and culture models of glaucoma for studying neuroprotection.

Li Y, Juul SE, Morris-Wiman JA, Calhoun DA, Christensen RD (1996) Erythropoietin receptors are expressed in the central nervous system of mid-trimester human fetuses. Pediatr Res 40:376-380.

Liu C, Shen K, Liu Z, Noguchi CT (1997) Regulated human erythropoietin receptor expression in mouse brain. J Biol Chem 272:32395-32400.

Liu ZY, Chin K, Noguchi CT (1994) Tissue specific expression of human erythropoietin receptor in transgenic mice. Dev Biol 166:159-169.

Marti HH, Wenger RH, Rivas LA, Straumann U, Digicaylioglu M, Henn V, Yonekawa Y, Bauer C, Gassmann M (1996) Erythropoietin gene expression in human, monkey and murine brain. Eur J Neurosci 8:666-676.

Masuda S, Okano M, Yamagishi K, Nagao M, Ueda M, Sasaki R (1994) A novel site of erythropoietin production. Oxygen-dependent production in cultured rat astrocytes. J Biol Chem 269:19488-19493.

Morishita E, Masuda S, Nagao M, Yasuda Y, Sasaki R (1997) Erythropoietin receptor is expressed in rat hippocampal and cerebral cortical neurons, and erythropoietin prevents in vitro glutamate-induced neuronal death. Neuroscience 76:105-116.

Pease ME, McKinnon SJ, Quigley HA, Kerrigan-Baumrind LA, Zack DJ (2000) Obstructed axonal transport of BDNF and its receptor TrkB in experimental glaucoma. Invest Ophthalmol Vis Sci 41:764-774.

Ratajczak J, Majka M, Kijowski J, Baj M, Pan ZK, Marquez LA, Janowska-Wieczorek A, Ratajczak MZ (2001) Biological significance of MAPK, AKT and JAK-STAT protein activation by various erythropoietic factors in normal human early erythroid cells. Br J Haematol 115:195-204.

Robinson MJ, Stippec SA, Goldsmith E, White MA, Cobb MH (1998) A constitutively active and nuclear form of the MAP kinase ERK2 is sufficient for neurite outgrowth and cell transformation. Curr Biol 8:1141-1150.

Sadamoto Y, Igase K, Sakanaka M, Sato K, Otsuka H, Sakaki S, Masuda S, Sasaki R (1998) Erythropoietin prevents place navigation disability and cortical infarction in rats with permanent occlusion of the middle cerebral artery. Biochem Biophys Res Commun 253:26-32.

Sakamoto H, Kitamura T, Yoshimura A (2000) Mitogen-activated protein kinase plays an essential role in the erythropoietin-dependent proliferation of CTLL-2 cells. J Biol Chem 275:35857-35862.

Sakanaka M, Wen TC, Matsuda S, Masuda S, Morishita E, Nagao M, Sasaki R (1998) In vivo evidence that erythropoietin protects neurons from ischemic damage. Proc Natl Acad Sci U S A 95:4635-4640.

Siren AL, Fratelli M, Brines M, Goemans C, Casagrande S, Lewczuk P, Keenan S, Gleiter C, Pasquali C, Capobianco A, Mennini T, Heumann R, Cerami A, Ehrenreich H, Ghezzi P (2001) Erythropoietin prevents neuronal apoptosis after cerebral ischemia and metabolic stress. Proc Natl Acad Sci U S A 98:4044-4049.

Socolovsky M, Nam H, Fleming MD, Haase VH, Brugnara C, Lodish HF (2001) Ineffective erythropoiesis in Stat5a(-/-)5b(-/-) mice due to decreased survival of early erythroblasts. Blood 98:3261-3273.

Uddin S, Kottegoda S, Stigger D, Platanias LC, Wickrema A (2000) Activation of the Akt/FKHRL1 pathway mediates the antiapoptotic effects of erythropoietin in primary human erythroid progenitors. Biochem Biophys Res Commun 275:16-19.

Weishaupt JH, Diem R, Kermer P, Krajewski S, Reed JC, Bähr M (2003a) Contribution of caspase-8 to apoptosis of axotomized rat retinal ganglion cells in vivo. Neurobiol Dis 13:124-135.

Weishaupt JH, Rohde G, Polking E, Siren AL, Ehrenreich H, Bähr M (2004) Effect of erythropoietin axotomy-induced apoptosis in rat retinal ganglion cells. Invest Ophthalmol Vis Sci 45:1514-1522.

Weishaupt JH, Kussmaul L, Grotsch P, Heckel A, Rohde G, Romig H, Bähr M, Gillardon F (2003b) Inhibition of CDK5 is protective in necrotic and apoptotic paradigms of neuronal cell death and prevents mitochondrial dysfunction. Mol Cell Neurosci 24:489-502.

Yoshimura A, Misawa H (1998) Physiology and function of the erythropoietin receptor. Curr Opin Hematol 5:171-176.

Chapter 8

ERYTHROPOIETIN FOR TREATMENT OF HUMAN BRAIN DISEASE: *EXPERIENCE FROM PROOF-OF-CONCEPT TRIALS*

Hannelore Ehrenreich and Anna-Leena Sirén
Clinical Neuroscience Laboratory, Max-Planck-Institute for Experimental Medicine, Göttingen, GERMANY

Abstract: Using a unique human "rat study" design we demonstrated for the first time that the hematopoietic growth factor erythropoietin (EPO) represents a successful and safe approach to pharmacological neuroprotection in human stroke. This approach is currently further explored and expanded in a multi-center setting. At the same time, EPO is being tested as neuroprotective add-on treatment in a chronic human brain disease, schizophrenia. Learning from previous mistakes, the design of such investigator initiated proof-of-principle studies has to be meticulous and thorough, scientifically sound and financially achievable. It should count on scientific enthusiasm and intrinsic motivation of involved investigators rather than on financial interests ("head-hunter approaches"). A constant "self-observing and -cleaning process" among study centers helps to guarantee a high level of quality. These factors will improve the success of future translational studies with EPO/EPO-analogues, such as the novel "designer-cytokine" CEPO (carbamylated erythropoietin), in a whole range of neurodegenerative diseases.

Key words erythropoietin; stroke; schizophrenia; neurodegeneration; neuroprotection

1. THE IDEA...

The availability of recombinant human erythropoietin (rhEPO) has revolutionized the treatment of patients suffering from anemia of chronic renal disease (Eschbach et al., 1987; Bauer and Kurtz, 1989; Jelkmann, 1992). Its introduction into the clinic in the mid to late eighties has revived the lives of millions of people. The early EPO literature alludes to improvement in cognitive function and quality of life upon chronic rhEPO treatment in uremic encephalopathy but attributes it to the increase in hematocrit caused by EPO (Di Paolo et al., 1992; Nissenson, 1992; Kramer et al., 1996; Pickett et al., 1999; Ehrenreich and Sirén, 2001b; Elwood et al., 2001). As early as 1987, however, one of us (HE) repeatedly observed a distinct improvement in cognitive performance of chronic renal failure patients treated with rhEPO before any measurable response in hematocrit. The most logical consequence out of this observation would have been to investigate the effect of EPO on brain functions. At this time, however, EPO was simply considered a renal hormone, stimulating red blood cell production. Ideas addressing its potential actions on the central nervous system were not appreciated. Retrospectively, many nephrologists all over the world admit of having had similar experiences when first treating patients suffering from chronic renal disease with rhEPO.

2. CONSOLIDATION...

Our original idea of EPO playing a role in the central nervous system had a virtual renaissance in the beginning of the nineties when EPO and its receptor were detected in the brain of various animal species and man (Masuda et al., 1993; Masuda et al., 1994; Marti et al., 1996; Juul et al., 1998; Schmitt et al., 1999). Why would EPO and EPOR be expressed in the brain? Would there be a similar induction of EPO production in brain cells as compared to the periphery? In other words, would hypoxia play a role in the pathophysiology of the brain EPO system? We first started to address these questions in cell culture experiments and in animal models of hypoxia/ischemia and neurotrauma (Lewczuk et al., 2000; Sirén and Ehrenreich, 2001; Sirén et al., 2001b; Knabe et al., 2004; Sättler et al., 2004; Weishaupt et al., 2004). When it became obvious that EPO has potent neuroprotective properties both in vitro and in vivo, we screened post mortem brain tissue from patients who had suffered a stroke, for the expression of EPO and EPO receptors. Strikingly, we observed a massive increase in EPO and EPO receptor immunoreactivity in neurons, astrocytes and endothelial cells of the infarcted brain (Sirén et al., 2001a). The remarkable upregulation of the cerebral EPO system in hypoxic/ischemic human brain together with the multitude of cytoprotective actions of EPO (see the other chapters of this book) encouraged us to plan a first therapeutic approach to human stroke using EPO.

3. STROKE: PLEA FOR EARLY HUMAN *PROOF-OF-PRINCIPLE* TRIALS

The pharmacological profile of the growth factor EPO in the nervous system makes it an ideal candidate for pharmacotherapeutic neuroprotection: EPO is anti-apoptotic, anti-oxidative, anti-inflammatory, anti-glutamergic, neurotrophic, and angiogenic, and modulates stem cell differentiation and proliferation (Tabira et al., 1995; Campana et al., 1998; Sakanaka et al., 1998; Bernaudin et al., 1999; Ribatti et al., 1999; Brines et al., 2000; Chattopadhyay et al, 2000; Digicaylioglu and Lipton, 2001; Kawakami et al., 2001; Shingo et al., 2001; Sirén and Ehrenreich, 2001; Sirén et al., 2001b; Genc et al., 2002; Grasso et al., 2002; Grimm et al., 2002; Ruscher et al., 2002; Springborg et al., 2002; Yu et al., 2002; Villa et al., 2003; Sättler et al., 2004; Weishaupt et al., 2004). For more detailed description of the pharmacological properties of EPO as derived from cell culture and animal studies please refer to the other chapters of this book. From a conceptual point of view, EPO therefore had to be tested for its efficacy in human stroke. Having a safe and well-tolerated compound on hand, we felt that there was no need to wait any longer for additional results from preclinical studies, especially, since the translation of numerous previous preclinical studies to man had yielded disappointing results, despite promising animal data. Basically all neuroprotective approaches to human stroke had failed thus far. Reasons for these failures are multitudinous and range from general problems in translating animal studies to man (species differences, dose and application, timing, animal model used), toxicity and intolerance reactions, to a tremendously high heterogeneity of the human population, particularly in studies recruiting unclearly defined patient groups (De Keyser et al., 1999; Davis et al., 2000; Albers et al., 2001; Green et al., 2003; Muir et al., 2004). These negative experiences in translating animal data to human studies motivated us to launch immediately a proof-of-concept trial using the clinically safe EPO for human stroke but, unfortunately, made it also very difficult to find industry partners spirited enough to adequately support this work. The frustration of pharmacological industry about failed trials and the resulting hesitation or unwillingness to finance any further trials has in fact accompanied our way of using EPO for human brain disease until now.

3.1 EPO for neuroprotection: Which dose/regimen to choose?

Our own encouraging preclinical data with EPO as a neuroprotective agent together with supporting evidence from other groups that had commenced to appear in the literature at around that time, laid the ground for our decision to initiate a proof-of-principle study in human stroke patients. We decided to start out with the highest EPO dose reported to be well tolerated in man at that time (Eschbach et al., 1987) and to apply it as an intravenous infusion. The relatively high dose should help to achieve EPO blood levels high enough to result in appreciable concentrations in the infarcted brain. The infusion over 30 min should allow to closely monitor potential side effects (e.g. blood pressure increases) and to stop the EPO application immediately if necessary.

Over the years, we as well as others were able to show that with respect to the nervous system, EPO has a bell-shaped dose response curve both in vitro and in vivo, for instance in its antiapoptotic effect (Sakanaka et al., 1998; Weishaupt et al., 2004). Although it is unclear whether this bell-shaped dose response will also hold true for the effects of EPO in human brain disease, this possibility has to be considered when designing large trials with various treatment arms in order to determine the most effective dose for a particular indication. In fact, we might have been fortunate to obtain positive results in our stroke trial with the dose used, not knowing at which point of the potentially bell-shaped curve we are.

3.2 Getting started: The safety profile of EPO in a new indication

Our application to the Ethical Committee in 1997 for the first trial using EPO in human stroke was able to build on the well-documented safety and tolerability of EPO over more than a decade of clinical use in patients suffering from anemia, particularly anemia related to chronic renal disease or to malignancies (Eschbach et al., 1987; Bohlius et al., 2002). In addition to these millions of patients, a considerable number of individuals had been treated with EPO in order to prepare for autologous blood transfusions during planned operations (Schlaeppi et al., 1994; Bohlius et al., 2002; Rosencher and Ozier, 2003). Our main argument for starting the trial at this early time point was therefore that, after a plethora of failed translational approaches to neuroprotection in human stroke (De Keyser et al., 1999; Davis et al., 2000; Albers et al., 2001; Green et al., 2003; Muir et al., 2004), there was no rationale for further delaying a proof-of-principle study in stroke patients using EPO, a safe and well tolerated compound. The question arose whether, despite the high safety profile of EPO in treatment of anemia, one would have to expect additional safety concerns in this new indication. Therefore, the first part of the Göttingen EPO stroke trial consisted of a safety study, with open-label application of EPO. Particular care was taken to monitor blood pressure in the stroke patients, considering the potential risk of blood pressure elevations as observed in subjects with renal failure upon chronic administration of EPO (Sundal et al., 1991). In addition, hematocrit and hemoglobin were determined over weeks after the application of EPO to stroke patients in order to control for potential increases in red blood cell counts that might cause rheological problems. None of these concerns turned out to be valid in stroke. There were no acute or delayed alterations in blood pressure upon infusion of EPO, and there was no increase in hematocrit following EPO treatment after stroke. In fact, we found that the EPO treated patients maintained their hematocrit whereas the control group showed a significant decrease in hematocrit over time after stroke (Ehrenreich et al., 2002).

3.3 The blood-brain-barrier: A barrier for EPO?

Another important issue of the safety trial was the question whether EPO, a large molecule (molecular weight over 30 000 Dalton) would have to be expected to cross the blood-brain-barrier and to reach the affected brain tissue and its environment in a sufficient concentration to exert its protective effects. To answer this question we had to determine the EPO concentrations in the cerebrospinal fluid (CSF) after application of EPO to stroke patients. As shown in Figure 1, there was indeed a 60-100 times over baseline accumulation of immunoreactive EPO in the CSF. Nevertheless, concentrations in the CSF do not really reflect the concentration of a compound in the brain. Molecules that are expected to be receptor-bound in brain tissue will be trapped there as soon as they cross the blood-brain-barrier. The CSF concentration, therefore, represents, at least in part, a "spillover" from brain tissue, drained into the CSF. It is well established that a breakdown of the blood-brain-barrier has to be expected after stroke (Ballabh et al., 2004). This fact, together with our finding of an increased density of EPOR expression in infarcted human brain tissue (Sirén et al., 2001a) and with the observation of a highly significant increase in EPO levels in the CSF (Ehrenreich et al., 2002), made it very likely that EPO would reach the infarcted brain tissue in sufficient amounts to exert its neuroprotective effects. Alleviating any remaining concerns, Brines and colleagues reported on animal data showing that EPO crosses the blood-brain-barrier to protect against experimental brain injury (Brines et al., 2000). Recently we demonstrated that EPO can reach the brain even via an intact blood-brain-barrier in both rodents and man: We found (1) elevated levels of EPO in the CSF of normal rats peaking at 3.5 hours after systemic administration of EPO, and (2) enrichment of In-111 rhEPO in the brain of healthy human individuals after intravenous bolus injection (see below and Ehrenreich et al., 2004).

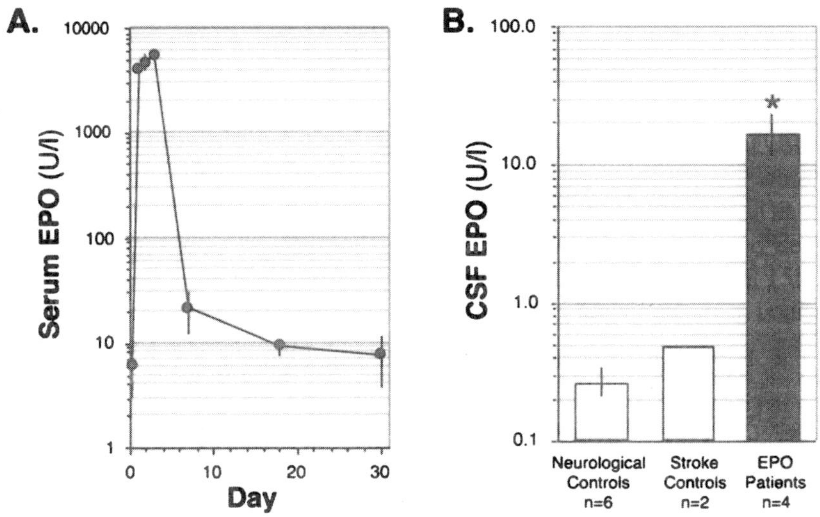

Figure 8-1. EPO concentrations in serum and cerebrospinal fluid. **A)** Serum EPO levels in stroke patients before EPO administration and 3h after EPO infusion on days 1, 2 and 3 after stroke. Data represent mean ± SEM. Note the logarithmic scaling of the ordinate. **B)** Cerebrospinal fluid EPO levels on day 2 in 4 stroke patients receiving EPO, in 6 control patients with other neurological conditions and in two stroke patients who did not receive EPO. Data represent mean ± SEM, *p<0.01 as compared to the neurological disease controls by paired independent t-test. Note the logarithmic scaling of the ordinate. (For further details see Ehrenreich et al., 2002).

3.4 Setting up a human "rat study"

An investigator initiated proof-of-principle study has to do what its name implies: prove the principle rather than count on a "robust effect" as sometimes requested by study designers in industry trials. When setting up the first EPO stroke trial, we decided to design a study that should come as close as possible to a preclinical animal study. When going from the established MCA (middle cerebral artery) occlusion models of laboratory animals to human stroke we are suddenly confronted with a tremendous heterogeneity of the test population as compared to the very homogeneous situation in the laboratory, and should therefore at least aim at defining a clear "lesion model" in man. Therefore, and also to make comparability of treatment groups easier, both clinically and with respect to imaging, we decided to restrict the stroke population in our trial to patients suffering from a stroke in the MCA territory. This was done anticipating that such clear-cut design would be at the cost of a much longer duration of the study.

A proof-of-principle trial, restricted to a relatively small number of patients, cannot afford to include patients without a clear diagnosis and a reliable time-window of inclusion after the onset of a stroke. It was therefore necessary and totally exceptional at that time to request magnetic resonance imaging (MRI) for diagnosis and as prerequisite to include a patient into the study. MRI comprising both diffusion weight and FLAIR imaging allowed us not only to secure the diagnosis but also to estimate the reliability of the time of symptom onset as derived from reports of patient and relatives. Whereas diffusion weight image should early show a clear lesion area, FLAIR imaging upon inclusion should be essentially normal. These two very unusual and restrictive inclusion criteria, stroke in the MCA territory and requirement of MRI for diagnosis, most likely explain the success of this small trial (Figure 2).

3.5 Efficacy: How to measure it for *proof-of-principle*

In contrast to large phase III trials with a clear-cut primary outcome, proof-of-principle trials should be set up to learn as much as possible about the potential efficacy of the compound to be evaluated. Therefore, not only the usual primary outcome measures of stroke (Barthel-Index, modified Rankin-Scale) were included in the follow-up investigations within the trial but also a careful examination of neurological symptoms after stroke (NIH Stroke Scale, Scandinavian Stroke Scale), an analysis of lesion size by MRI on days 1, 3 and 18 after stroke, as well as regular determinations of circulating damage markers (S100B). Based on the results of this careful follow-up evaluation, primary outcome measures could then be determined for larger phase III trials.

Inclusion Criteria

+

- **Ischemic stroke in the MCA territory**
- **confirmed by cMRI (diffusion weighted)**
- Age ≤ 80 years
- Treatment within 8h after onset of symptoms
- Informed consent of patient or relatives

−

- MRI contraindications
- Onset of symptoms unclear
- Fast resolving neurologic deficit
- Coma or pre-coma
- Intracranial hemorrhage, neoplasia, septic embolism
- Malignant hypertension
- Renal failure
- Myeloproliferative disorders
- Allergy against erythropoietin
- Participation in other trials

Figure 8-2. Inclusion and exclusion criteria of the "Göttingen EPO Stroke Trial". (For further details see Ehrenreich et al., 2002).

3.6 The "Göttingen EPO stroke trial"

After successful conclusion of the safety part of the EPO stroke trial, a double blind proof-of-concept trial was started including a total of 40 patients. The design of the trial remained identical to that of the safety study (Figure 3). We found that patients treated with EPO had a better outcome as compared to placebo treated control patients: (1) They had a significantly less prominent neurological deficit in the early post-stroke situation. (2) They had a better restitution of brain function as determined by the outcome scales, Barthel Index and modified Rankin Scale. (3) Evolution of lesion size in the EPO treated patients was significantly smaller as compared to controls. (4) Circulating levels of the glial damage marker S100B were not only significantly lower in the EPO as compared to the placebo group but also showed a faster return to baseline levels.

Complimentary to clinical evaluation and imaging, the use of circulating biomarkers of brain injury will be of growing importance for a developing field of neuroprotection in human brain disease (Herrmann and Ehrenreich, 2003). In fact, the "Göttingen EPO-stroke trial" is the first therapeutic study showing a clear difference in plasma levels of a glial damage marker between treatment groups. Much work will have to go into future development of analytical panels of biomarkers that will allow follow-up and estimation of success of therapeutic approaches to various, also chronic, brain diseases (Herrmann and Ehrenreich, 2003).

The fact that a merely neuroprotective study showed beneficial effects in the treatment of human stroke should encourage further trials aiming at neuroprotection in these conditions.

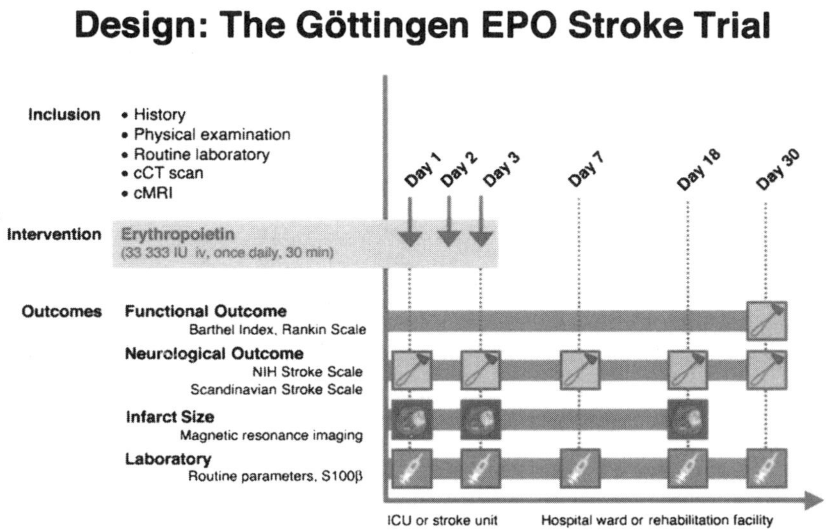

Figure 8-3. Study design of the "Göttingen EPO Stroke Trial". (For further details see Ehrenreich et al., 2002).

3.7 Multicenter trials in human brain disease: Learning from mistakes and how to save money

Since the monetary support of investigator-initiated trials is usually not comparable with the ample funding of industry driven studies, the design of such trials has to be extremely careful, scientifically sound and financially achievable. One method of saving a tremendous amount of money and increasing the trial quality is to abstain from so-called "head-hunter approaches".

Rather than paying a center for each included patient (which harbors the danger that the inclusion-criteria are violated), our multi-center trial counts on scientific enthusiasm and intrinsic motivation of the involved investigators. In addition to the leaders of the participating departments, a responsible person "in the front" is identified who is actually paid by the trial funds and is directly responsible for recruitment efficiency, data collection and transfer to the master center. A constant "self-observing and -cleaning process" among the collaborating centers will also guarantee a high level of quality and may cause one or the other center to fall out along the way...

3.8 The German EPO stroke multicenter trial

Based on the successful proof-of-principle trial on EPO in stroke, a multicenter trial has been designed and started in February 2003. For the design of this trial it was important that neuroprotection with EPO in stroke is not meant to compete with the use of rTPA (recombinant tissue plasminogen activator), the only available specific treatment for stroke, or related compounds used for thrombolytic therapy (The National Institute of Neurological Disorders and Stroke rt-PA Stroke Study Group, 1995; Hacke et al., 1998). On the contrary, the trial was set up to test EPO as an addition to rTPA treatment, targeting a different set of pathophysiological mechanisms. Provided that the success of our small proof-of-principle trial can be repeated, EPO will be complimentary to rTPA therapy, i.e. be used in combination with rTPA or serve as an only alternative for patients who are excluded from the use of rTPA. Up to date, such alternative therapy would apply for over 90% of stroke patients with contraindications to rTPA treatment.

At present a multicenter study "EPO in stroke" is being carried out in Germany, with Bremen, Göttingen and Hanover as participating centers and a planned expansion to a total of eight centers. The study is oriented on the successful pilot study; it does, however, include some novel features. For one, only patients with moderate to severe infarcts are included (NIH Stroke Scale score > 5). The time window for the first application of EPO comprises 6 hours instead of 8 hours, the application of rTPA, as outlined above, is allowed (stratification for treatment and center), the period of follow-up examinations stretches out to 90 days (primary outcome: Barthel Index on day 90) and even a year. The imaging controls are reduced to two sessions (days 1 and 7). Instead of 3 x 33.333 units of EPO-ß, the patients now receive 3 x 40.000 units of EPO-α. A first safety interim analysis in March 2004 found EPO to be safe and well tolerated in stroke patients with and without accompanying rTPA treatment. No data are as yet available with respect to efficacy of this multicenter trial. Should this trial be able to reproduce the success of the first proof-of-concept study in a total of over 500 patients, then EPO may be used in the future as a drug that can easily be applied already in the emergency car.

4. SCHIZOPHRENIA: THE "PROTOTYPE" CHRONIC INDICATION FOR NEUROPROTECTION

Taking into account the neuroprotective properties of EPO in preclinical studies, both in vitro and in vivo (Tabira et al., 1995; Campana et al., 1998; Sakanaka et al., 1998; Bernaudin et al., 1999; Ribatti et al., 1999; Brines et al., 2000; Chattopadhyay et al., 2000; Digicaylioglu and Lipton, 2001; Kawakami et al., 2001; Shingo et al., 2001; Sirén and Ehrenreich, 2001; Sirén et al., 2001b; Genc et al., 2002; Grasso et al., 2002; Grimm et al., 2002; Ruscher et al., 2002; Springborg et al., 2002; Yu et al., 2002; Villa et al., 2003; Sättler et al., 2004; Weishaupt et al., 2004), one might argue that by targeting pathophysiological mechanisms that belong to a final common deleterious pathway in a whole range of neurodegenerative diseases, EPO would be a candidate for proof-of-concept therapeutic trials in diseases such as Alzheimer's disease, Parkinson's disease, Multiple Sclerosis, diabetic neuropathy, spinal cord injury, to name a few. Why did we choose schizophrenia as the first proof-of-concept trial on the chronic use of EPO in neurodegenerative disease? The potential target population affected by this disease is a huge one. Estimated incidence of schizophrenia amounts to 1-3 % of the population, predominantly young people, and causes a tremendous economic burden to society in addition to the formidable consequences for each affected individual and the suffering of his environment. Also, neuroprotection is a completely new approach to the treatment of schizophrenia (Ehrenreich and Sirén, 2001a). Over the last decade it became more and more evident that in addition to disturbances of brain development, continuous neurodegenerative processes are involved in the pathogenesis of this disease (Moldin and Gottesman, 1997; Lieberman, 1999; Andreasen, 2000; Benes, 2000; Young et al., 2000; Malaspina et al., 2001; Lewis and Levitt, 2002; Prabakaran et al., 2004). Nevertheless, molecular and cellular mechanisms accounting for the pathology of schizophrenic psychosis are far from clear. Existing animal models of schizophrenia are only able to represent certain aspects of this often devastating human disease (Lipska and Weinberger, 2000; Kilts, 2001).

Therefore, novel concepts of treating schizophrenia should be translated into clinical use in the form of pilot trials without any delay; even more so, if the suggested beneficial compounds in this disorder are safe. From successful human trials, there ought to be a way back to animal research further explaining the mechanisms of action and, perhaps, helping to improve the dose-finding mode and ways and intervals of application.

We know that during the episodes of schizophrenic psychosis a dramatic worsening of cognitive performance can be observed in many cases (Lieberman, 1999; Haefner, 2000; Kremen et al., 2001). More than hundred years ago, Emil Kraepelin described these phenomena and called the disease "dementia praecox" (Kraepelin, 1919). Antipsychotic therapy, one of the major achievements of psychiatry of the last century, has prepared the ground for any further therapy aiming at influencing the course of the disease (Carlsson, 1988). Reduction in psychotic symptoms through antipsychotic treatment is a prerequisite for creating a stable therapeutic interaction between patients and therapists. Antipsychotic treatment by itself, however, is unlikely to influence the neurodegenerative course of schizophrenia. Neurodegenerative processes might even be aggravated during chronic antipsychotic treatment due to the potential neurotoxicity of classical neuroleptic drugs, an effect that EPO is capable of suppressing (Ehrenreich et al., 2004). Therefore, a first neuroprotective add-on therapy in schizophrenia using EPO as a well-tolerated and safe compound appeared worthwhile performing in man.

4.1 EPO for schizophrenia: Preparing for a *proof-of-principle* trial

In 1999, we decided to use EPO as a first compound with known high neuroprotective potential to design a proof-of-concept trial "Neuroprotection in schizophrenia". Four key questions had to be clarified before starting the clinical trial: (1) Can EPO, in a disease with predominantly intact blood-brain barrier, reach the brain in an appreciable amount following intravenous application? (2) Would we be able to detect EPO receptors as prerequisites for its action in the brain of schizophrenic patients? (3) Can chronic EPO application be considered safe in this indication? (4) Can we expect any benefit for cognitive functioning by EPO treatment in schizophrenic patients? In preparation of this first neuroprotective add-on trial in schizophrenia we were able to address each of these key points successfully (Ehrenreich et al., 2004).

4.2 German Multicenter trial: EPO as neuroprotective add-on therapy in chronic schizophrenic patients

Based on the results summarized above, the first multi-center trial in schizophrenia was started in Germany in April 2003. Results of this trial should be available beginning of the year 2005. This trial aims at improvement of cognitive function in chronic schizophrenics with a clear-cut cognitive deficit upon inclusion. Although a trial including acutely ill schizophrenic patients, preferably those experiencing their first episode, would theoretically have been more appropriate for proof-of-principle, we decided to go for a trial in chronic schizophrenic patients. Reasons for this decision include: (1) Better management of chronically ill patients who are familiar with the nature of their disease. (This helps in reducing the probability of dropouts.) (2) The possibility to get a clear-cut informed consent. (This is very often difficult in acutely ill schizophrenic patients.) (3) The need to have a clearly defined cognitive deficit as inclusion criterion. (This is basically impossible to reliably judge in an acutely ill patient.) (4) The necessity to recruit a small, well-defined patient group and to avoid any diagnostic failures (Schizophrenia is not a cross-sectional diagnosis). In our understanding, neuroregenerative processes are constantly taking place even in a diseased brain. Should EPO be able to enhance this neurorestoration; then we would be successful in proving cognitive amelioration even in chronically ill patients. With this first neuroprotective add-on trial in schizophrenia, EPO has been introduced as an example of a neuroprotective strategy for a chronic brain disease.

5. CONCLUSIONS: EPO, CEPO, AND OTHER EPO-ANALOGUES; A BOOST FOR TRANSLATIONAL NEUROSCIENCE... *WHERE TO GO FROM HERE?*

EPO represents for the first time a successful and safe approach to pharmacological neuroprotection in human brain disease. This approach has to be further explored, improved and expanded. Animal studies will continue to illuminate on mechanisms of action. Combinations of EPO/EPO-analogues with other neuroprotective compounds as well as with neuropsychological and physiotherapeutic measures will be investigated both in preclinical and clinical studies. The novel "designer-cytokine" CEPO (carbamylated erythropoietin) maintains the neuroprotective action profile of EPO but is devoid of its hematopoietic effects providing a major step forward in translational neuroscience (Ehrenreich, 2004; Leist et al., 2004). As soon as safety-trials have been concluded, it will be worthwhile to test CEPO for a whole range of neurodegenerative diseases requiring chronic application. In contrast to EPO with its potent hematopoietic effects and therefore an increased risk of thromboembolic complications in chronic treatment, CEPO will be a tremendous improvement in these indications (Ehrenreich, 2004; Leist et al., 2004). Other, similar compounds are presently being developed. In addition, combination therapies with other growth factors may prove useful. For instance, a combined administration of EPO with insulin-like growth factor I (IGF-I) has been reported most recently to exert synergistic neuroprotective effects (Digicaylioglu et al., 2004) that may represent a promising extension of the neuroprotective profile of EPO in man. Therapeutic neuroprotection in conditions like neurotrauma, spinal cord injury, Parkinson, Alzheimer or multiple sclerosis is currently under planning as is a potential prophylactic use of EPO for neuroprotective strategies in high risk patient groups (for instance radio- and/or chemotherapy, or hepatic encephalopathy) and as an adjuvant application to extensive rehabilitation programs after stroke or neuroinjury.

ACKNOWLEDGEMENTS

The clinical studies reported here have been funded by the Max-Planck-Society (MPG), and by research grants from Orthobiotech and Lundbeck. The basic research has received grant support from MPG, DFG, Orthobiotech, Lundbeck, and Roche.

REFERENCES

Albers GW, Goldstein LB, Hall D, Lesko LM (2001) Aptiganel hydrochloride in acute ischemic stroke: a randomized controlled trial. Jama 286:2673-2682.

Andreasen NC (2000) Schizophrenia: the fundamental questions. Brain Res Brain Res Rev 31:106-112.

Ballabh P, Braun A, Nedergaard M (2004) The blood-brain barrier: an overview: structure, regulation, and clinical implications. Neurobiol Dis 16:1-13.

Bauer C, Kurtz A (1989) Oxygen sensing in the kidney and its relation to erythropoietin production. Annu Rev Physiol 51:845-856.

Benes FM (2000) Emerging principles of altered neural circuitry in schizophrenia. Brain Res Brain Res Rev 31:251-269.

Bernaudin M, Marti HH, Roussel S, Divoux D, Nouvelot A, MacKenzie ET, Petit E (1999) A potential role for erythropoietin in focal permanent cerebral ischemia in mice. J Cereb Blood Flow Metab 19:643-651.

Bohlius J, Langensiepen S, Schwarzer G, Engert A (2002) Epoetin in the treatment of malignant disease: A comprehensive meta-analysis. Blood 100:3430A.

Brines ML, Ghezzi P, Keenan S, Agnello D, de Lanerolle NC, Cerami C, Itri LM, Cerami A (2000) Erythropoietin crosses the blood-brain barrier to protect against experimental brain injury. Proc Natl Acad Sci U S A 97:10526-10531.

Campana WM, Misasi R, O'Brien JS (1998) Identification of a neurotrophic sequence in erythropoietin. Int J Mol Med 1:235-241.

Carlsson A (1988) The current status of the dopamine hypothesis of schizophrenia. Neuropsychopharmacology 1:179-186.

Chattopadhyay A, Choudhury TD, Bandyopadhyay D, Datta AG (2000) Protective effect of erythropoietin on the oxidative damage of erythrocyte membrane by hydroxyl radical. Biochem Pharmacol 59:419-425.

Davis SM, Lees KR, Albers GW, Diener HC, Markabi S, Karlsson G, Norris J (2000) Selfotel in acute ischemic stroke : possible neurotoxic effects of an NMDA antagonist. Stroke 31:347-354.

De Keyser J, Sulter G, Luiten PG (1999) Clinical trials with neuroprotective drugs in acute ischaemic stroke: are we doing the right thing? Trends Neurosci 22:535-540.

Di Paolo B, Di Liberato L, Fiederling B, Catucci G, Bucciarelli S, Paolantonio L, Albertazzi A (1992) Effects of uremia and dialysis on brain electrophysiology after recombinant erythropoietin treatment. Asaio J 38:M477-480.

Digicaylioglu M, Lipton SA (2001) Erythropoietin-mediated neuroprotection involves cross-talk between Jak2 and NF-kappaB signalling cascades. Nature 412:641-647.

Digicaylioglu M, Garden G, Timberlake S, Fletcher L, Lipton SA (2004) Acute neuroprotective synergy of erythropoietin and insulin-like growth factor I. Proc Natl Acad Sci U S A 101:9855-9860.

Ehrenreich H (2004) A boost for translational neuroscience. Science 305:184-185.

Ehrenreich H, Sirén A-L (2001a) Special issue-editorial: Neuroprotection - what does it mean? - What means do we have? European Archives of Psychiatry and Clinical Neuroscience 251:149-151.

Ehrenreich H, Sirén A-L (2001b) Benefits of recombinant human erythropoietin on cognitive function. Erythropoiesis 11:35-40.

Ehrenreich H, Degner D, Meller J, Brines M, Behe M, Hasselblatt M, Woldt H, Falkai P, Knerlich F, Jacob S, von Ahsen N, Maier W, Bruck W, Ruther E, Cerami A, Becker W, Sirén A-L (2004) Erythropoietin: a candidate compound for neuroprotection in schizophrenia. Mol Psychiatry 9:42-54.

Ehrenreich H, Hasselblatt M, Dembowski C, Cepek L, Lewczuk P, Stiefel M, Rustenbeck H-H, Breiter N, Jacob S, Knerlich F, Bohn M, Poser W, Rüther E, Kochen M, Gefeller O, Gleiter C, Wessel TC, De Ryck M, Itri L, Prange H, Cerami A, Brines M, Sirén A-L (2002) Erythropoietin therapy for acute stroke is both safe and beneficial. Molecular Medicine 8:495-505.

Elwood PC, Pickering J, Gallacher JE (2001) Cognitive function and blood rheology: results from the Caerphilly cohort of older men. Age Ageing 30:135-139.

Eschbach JW, Egrie JC, Downing MR, Browne JK, Adamson JW (1987) Correction of the anemia of end-stage renal disease with recombinant human erythropoietin. Results of a combined phase I and II clinical trial. N Engl J Med 316:73-78.

Genc S, Akhisaroglu M, Kuralay F, Genc K (2002) Erythropoietin restores glutathione peroxidase activity in 1-methyl-4-phenyl-1,2,5,6-tetrahydropyridine-induced neurotoxicity in C57BL mice and stimulates murine astroglial glutathione peroxidase production in vitro. Neurosci Lett 321:73-76.

Grasso G, Buemi M, Alafaci C, Sfacteria A, Passalacqua M, Sturiale A, Calapai G, De Vico G, Piedimonte G, Salpietro FM, Tomasello F (2002) Beneficial effects of systemic administration of recombinant human erythropoietin in rabbits subjected to subarachnoid hemorrhage. Proc Natl Acad Sci U S A 99:5627-5631.

Green RA, Odergren T, Ashwood T (2003) Animal models of stroke: do they have value for discovering neuroprotective agents? Trends Pharmacol Sci 24:402-408.

Grimm C, Wenzel A, Groszer M, Mayser H, Seeliger M, Samardzija M, Bauer C, Gassmann M, Reme CE (2002) HIF-1-induced erythropoietin in the hypoxic retina protects against light-induced retinal degeneration. Nat Med 8:718-724.

Hacke W, Kaste M, Fieschi C, von Kummer R, Davalos A, Meier D, Larrue V, Bluhmki E, Davis S, Donnan G, Schneider D, Diez-Tejedor E, Trouillas P (1998) Randomised double-blind placebo-controlled trial of thrombolytic therapy with intravenous alteplase in acute ischaemic stroke (ECASS II). Second European-Australasian Acute Stroke Study Investigators. Lancet 352:1245-1251.

Haefner H (2000) Epidemiology of schizophrenia. A thriving discipline at the turn of the century. Eur Arch Psychiatry Clin Neurosci 250:271-273.

Herrmann M, Ehrenreich H (2003) Brain derived proteins as markers of acute stroke: Their relation to pathophysiology, outcome prediction and neuroprotective drug monitoring. Restor Neurol Neurosci 21:177-190.

Jelkmann W (1992) Erythropoietin: structure, control of production, and function. Physiol Rev 72:449-489.

Juul SE, Anderson DK, Li Y, Christensen RD (1998) Erythropoietin and erythropoietin receptor in the developing human central nervous system. Pediatr Res 43:40-49.

Kawakami M, Sekiguchi M, Sato K, Kozaki S, Takahashi M (2001) Erythropoietin receptor-mediated inhibition of exocytotic glutamate release confers neuroprotection during chemical ischemia. J Biol Chem 276:39469-39475.

Kilts CD (2001) The changing roles and targets for animal models of schizophrenia. Biol Psychiatry 50:845-855.

Knabe W, Knerlich F, Washausen S, Kietzmann T, Sirén AL, Brunnett G, Kuhn HJ, Ehrenreich H (2004) Expression patterns of erythropoietin and its receptor in the developing midbrain. Anat Embryol (Berl) 207:503-512.

Kraepelin E (1919) Dementia praecox and paraphrenia. Edinburgh, Scotland: E & S Livingstone.

Kramer L, Madl C, Stockenhuber F, Yeganehfar W, Eisenhuber E, Derfler K, Lenz K, Schneider B, Grimm G (1996) Beneficial effect of renal transplantation on cognitive brain function. Kidney Int 49:833-838.

Kremen WS, Seidman LJ, Faraone SV, Tsuang MT (2001) Intelligence quotient and neuropsychological profiles in patients with schizophrenia and in normal volunteers. Biol Psychiatry 50:453-462.

Leist M, Ghezzi P, Grasso G, Bianchi R, Villa P, Fratelli M, Savino C, Bianchi M, Nielsen J, Gerwien J, Kallunki P, Larsen AK, Helboe L, Christensen S, Pedersen LO, Nielsen M, Torup L, Sager T, Sfacteria A, Erbayraktar S, Erbayraktar Z, Gokmen N, Yilmaz O, Cerami-Hand C, Xie QW, Coleman T, Cerami A, Brines M (2004) Derivatives of erythropoietin that are tissue protective but not erythropoietic. Science 305:239-242.

Lewczuk P, Hasselblatt M, Kamrowski-Kruck H, Heyer A, Unzicker C, Sirén AL, Ehrenreich H (2000) Survival of hippocampal neurons in culture upon hypoxia: effect of erythropoietin. Neuroreport 11:3485-3488.

Lewis DA, Levitt P (2002) Schizophrenia as a disorder of neurodevelopment. Annu Rev Neurosci 25:409-432.

Lieberman JA (1999) Is schizophrenia a neurodegenerative disorder? A clinical and neurobiological perspective. Biol Psychiatry 46:729-739.

Lipska BK, Weinberger DR (2000) To model a psychiatric disorder in animals: schizophrenia as a reality test. Neuropsychopharmacology 23:223-239.

Malaspina D, Goetz RR, Friedman JH, Kaufmann CA, Faraone SV, Tsuang M, Cloninger CR, Nurnberger JI, Jr., Blehar MC (2001) Traumatic brain injury and schizophrenia in members of schizophrenia and bipolar disorder pedigrees. Am J Psychiatry 158:440-446.

Marti HH, Wenger RH, Rivas LA, Straumann U, Digicaylioglu M, Henn V, Yonekawa Y, Bauer C, Gassmann M (1996) Erythropoietin gene expression in human, monkey and murine brain. Eur J Neurosci 8:666-676.

Masuda S, Okano M, Yamagishi K, Nagao M, Ueda M, Sasaki R (1994) A novel site of erythropoietin production. Oxygen-dependent production in cultured rat astrocytes. J Biol Chem 269:19488-19493.

Masuda S, Nagao M, Takahata K, Konishi Y, Gallyas F, Jr., Tabira T, Sasaki R (1993) Functional erythropoietin receptor of the cells with neural characteristics. Comparison with receptor properties of erythroid cells. J Biol Chem 268:11208-11216.

Moldin SO, Gottesman, II (1997) At issue: genes, experience, and chance in schizophrenia--positioning for the 21st century. Schizophr Bull 23:547-561.

Muir KW, Lees KR, Ford I, Davis S (2004) Magnesium for acute stroke (Intravenous Magnesium Efficacy in Stroke trial): randomised controlled trial. Lancet 363:439-445.

Nissenson AR (1992) Epoetin and cognitive function. Am J Kidney Dis 20:21-24.

Pickett JL, Theberge DC, Brown WS, Schweitzer SU, Nissenson AR (1999) Normalizing hematocrit in dialysis patients improves brain function. Am J Kidney Dis 33:1122-1130.

Prabakaran S, Swatton JE, Ryan MM, Huffaker SJ, Huang JT, Griffin JL, Wayland M, Freeman T, Dudbridge F, Lilley KS, Karp NA, Hester S, Tkachev D, Mimmack ML, Yolken RH, Webster MJ, Torrey EF, Bahn S (2004) Mitochondrial dysfunction in schizophrenia: evidence for compromised brain metabolism and oxidative stress. Mol Psychiatry 9:643.

Ribatti D, Presta M, Vacca A, Ria R, Giuliani R, Dell'Era P, Nico B, Roncali L, Dammacco F (1999) Human erythropoietin induces a pro-angiogenic phenotype in cultured endothelial cells and stimulates neovascularization in vivo. Blood 93:2627-2636.

Rosencher N, Ozier Y (2003) Peri-operative use of EPO. Transfus Clin Biol 10:159-164.

Ruscher K, Freyer D, Karsch M, Isaev N, Megow D, Sawitzki B, Priller J, Dirnagl U, Meisel A (2002) Erythropoietin is a paracrine mediator of ischemic tolerance in the brain: evidence from an in vitro model. J Neurosci 22:10291-10301.

Sakanaka M, Wen TC, Matsuda S, Masuda S, Morishita E, Nagao M, Sasaki R (1998) In vivo evidence that erythropoietin protects neurons from ischemic damage. Proc Natl Acad Sci U S A 95:4635-4640.

Sättler MB, Merkler D, Maier K, Stadelmann C, Ehrenreich H, Bähr M, Diem R (2004) Neuroprotective effects and intracellular signaling pathways of erythropoietin in a rat model of multiple sclerosis. Cell Death and Differentiation 11:S181-192.

Schlaeppi B, Gunter P, Nydegger UE (1994) Enhancing the efficacy of preoperative autologous blood donation by erythropoietin. Transfus Sci 15:171-177.

Schmitt M, Gleiter CH, Nichol JL, Pralle L, Hasselblatt M, Poser W, Ehrenreich H (1999) Haematological abnormalities in early abstinent alcoholics are closely associated with alterations in thrombopoietin and erythropoietin serum profiles. Thromb Haemost 82:1422-1427.

Shingo T, Sorokan ST, Shimazaki T, Weiss S (2001) Erythropoietin regulates the in vitro and in vivo production of neuronal progenitors by mammalian forebrain neural stem cells. J Neurosci 21:9733-9743.

Sirén A-L, Ehrenreich H (2001) Erythropoietin - a novel concept of neuroprotection. European Archives of Psychiatry and Clinical Neuroscience 251:179-184.

Sirén A-L, Knerlich F, Poser W, Gleiter C, Brück W, Ehrenreich H (2001a) Erythropoietin and erythropoietin receptor in human ischemic/hypoxic brain. Acta Neuropathologica 101:271-276.

Sirén AL, Fratelli M, Brines M, Goemans C, Casagrande S, Lewczuk P, Keenan S, Gleiter C, Pasquali C, Capobianco A, Mennini T, Heumann R, Cerami A, Ehrenreich H, Ghezzi P (2001b) Erythropoietin prevents neuronal apoptosis after cerebral ischemia and metabolic stress. Proc Natl Acad Sci U S A 98:4044-4049.

Springborg JB, Ma X, Rochat P, Knudsen GM, Amtorp O, Paulson OB, Juhler M, Olsen NV (2002) A single subcutaneous bolus of erythropoietin normalizes cerebral blood flow autoregulation after subarachnoid haemorrhage in rats. Br J Pharmacol 135:823-829.

Sundal E, Businger J, Kappeler A (1991) Treatment of transfusion-dependent anaemia of chronic renal failure with recombinant human erythropoietin. A European multicentre study in 142 patients to define dose regimen and safety profile. Nephrol Dial Transplant 6:955-965.

Tabira T, Konishi Y, Gallyas F, Jr. (1995) Neurotrophic effect of hematopoietic cytokines on cholinergic and other neurons in vitro. Int J Dev Neurosci 13:241-252.

The National Institute of Neurological Disorders and Stroke rt-PA Stroke Study Group (1995) Tissue plasminogen activator for acute ischemic stroke. N Engl J Med 333:1581-1587.

Villa P, Bigini P, Mennini T, Agnello D, Laragione T, Cagnotto A, Viviani B, Marinovich M, Cerami A, Coleman TR, Brines M, Ghezzi P (2003) Erythropoietin selectively attenuates cytokine production and inflammation in cerebral ischemia by targeting neuronal apoptosis. J Exp Med 198:971-975.

Weishaupt JH, Rohde G, Polking E, Sirén AL, Ehrenreich H, Bähr M (2004) Effect of erythropoietin axotomy-induced apoptosis in rat retinal ganglion cells. Invest Ophthalmol Vis Sci 45:1514-1522.

Young KA, Manaye KF, Liang C, Hicks PB, German DC (2000) Reduced number of mediodorsal and anterior thalamic neurons in schizophrenia. Biol Psychiatry 47:944-953.

Yu X, Shacka JJ, Eells JB, Suarez-Quian C, Przygodzki RM, Beleslin-Cokic B, Lin CS, Nikodem VM, Hempstead B, Flanders KC, Costantini F, Noguchi CT (2002) Erythropoietin receptor signalling is required for normal brain development. Development 129:505-516.

Chapter 9

ERYTHROPOIETIN IN SPINAL CORD INJURY

Challenges for a novel neuroprotective strategy

Michael Brines and Anthony Cerami
The Kenneth S. Warren Institute, 712 Kitchawan Road, Ossining, New York 10562

Abstract: Spinal cord injury (SCI) is a devastating condition lacking a clearly effective pharmacological treatment. The cytokine erythropoietin (EPO), which mediates cytoprotection in a variety of tissues through activation of multiple signaling pathways, is markedly effective in preclinical models of ischemic, traumatic and inflammatory SCI. The recent development of non-erythropoietic derivatives of EPO with outstanding preclinical characteristics encourages evaluation of tissue-protective cytokines in clinical trials of spinal cord injury.

Key words: spinal cord injury, erythropoietin, tissue protection, methylprednisolone, ischemia, inflammation, apoptosis, regeneration, animal models, neuroprotection, clinical trials

1. INTRODUCTION

Each year 11,000 new cases of spinal cord injury (SCI) are reported in the United States, over half of which occur among individuals under 30 years of age (National Spinal Cord Injury Statistical Center 2003). Aside from the incalculable human suffering, the medical surgical and rehabilitative care for patients with SCI is estimated at over four billion dollars per year (Kwon et al. 2002). Currently, there is no clearly effective therapy for spinal cord injury. Evidence supporting improved neurological recovery following treatment with the synthetic glucocorticosteroid methylprednisolone sodium succinate (MPSS), the standard therapeutic intervention for SCI, has been called into question in recent years. More importantly, this treatment may have potentially deleterious effects on early mortality and morbidity (Short et al. 2000; American Association of Neurological Surgeons 2002).

A number of other pharmacological means of reducing the extent of the injury are in various stages of animal and human evaluation (Blight et al. 2001). One of the most promising candidates is erythropoietin (EPO), a hematopoietic growth factor produced mainly by kidney and fetal liver, which stimulates proliferation and differentiation of erythroid precursor cell (Fisher 2003). However, EPO is also produced in the central nervous system (Sasaki 2003) and mediates neuroprotection against experimental brain injury and ischemia (Sakanaka et al. 1998; Brines et al. 2000; Digicaylioglu et al. 2001; Siren et al. 2001; Sasaki 2003). Notably, a recent clinical trial demonstrated significant improvement in outcome of stroke patients with documented non-hemorrhagic infarcts within the distribution of the middle cerebral artery who were given recombinant human EPO (rhEPO) intravenously within 8 hours of the onset of symptoms (Ehrenreich et al. 2002) .

Spinal cord injury shares many pathophysiological features with brain injury and recent studies in animal models indicate that EPO is very effective in attenuating the severity of spinal cord damage (Celik et al. 2002; Gorio et al. 2002; Sekiguchi et al. 2003; Kaptanoglu et al. 2004). This review discusses the results of these studies and the prospects for developing non-hematopoietic EPO-based drugs with an enhanced pharmacological profile for use in the clinic.

2. SPINAL CORD INJURY

Spinal cord injury is the result of damage to the nerves within the spinal canal, disrupting the spinal cord's ability to integrate afferent and efferent information for control of sensory, motor and autonomic functions. Injury commonly occurs during blunt, non-penetrating trauma (for example, as a result of vehicular accidents) whereby the bony and ligamentous components of the spinal column are laterally displaced and impart sudden compression to the spinal cord.

The trauma that occurs at one site within the spinal cord may kill many resident neurons, glia and capillary endothelial cells outright, but the clinical severity of the injury is dictated primarily by the fibers passing through the vicinity of the lesion that become non-functional because of edema and other factors. A cascade of more extensive cell death, thought to be caused by a combination of inflammation, ischemia and apoptosis (Dumont et al. 2001; Carlson et al. 2002; Norenberg et al. 2004), spreads out along the spinal cord away from the lesion epicenter as a function of time (hours to days), similar to a fire spreading outward from its ignition point. The extent of this secondary damage is what largely determines clinical outcome.

2.1 Animal models of SCI

The availability of animal models of SCI has facilitated an understanding of disease pathogenesis, as well as the evaluation of potential therapies. There are three general types of rodent SCI models commonly used: transection, compression and contusion (Rosenzweig et al. 2004). Transection involves cutting some of the spinal cord. As most humans suffer an acute blow to the spinal cord followed by a period of spinal cord compression, transection models do not mimic human injury well. For this reason, many researchers use compression and contusion models to evaluate therapeutic options. Compressive injuries are induced in rodents by squeezing the spinal cord with a modified aneurysm clip, forceps, or by placing a weight on the exposed spinal cord. Contusion injuries are developed by hitting the exposed spinal cord, most often with a dropped weight or a computerized solenoid-driven device to displace the cord by a precise amount.

Another, less severe type of injury is induced by crushing one of the nerve roots along the spinal cord (Kawakami et al. 1994). Lumbar disc prolapse and the ensuing compression of nerve roots is the most common cause of lower back pain. In rodent models this kind of injury can be induced by crushing nerve roots in the middle of the spine, in the lumbar region, using forceps. As a result of this injury, animals experience allodynia (i.e., a normal, innocuous stimulus is misinterpreted as pain) (Kawakami et al. 1994; Sekiguchi et al. 2003) .

In addition to trauma, there are other causes of SCI. For example, injury can be caused by temporary or permanent block in the spinal cord blood flow secondary during ischemia as a result of clamping the aorta in heart surgery (Gharagozloo et al. 1996). The primary targets of damage from the ensuing ischemia are the motor neurons in the ventral horn of the spinal cord. This type of injury can be modeled in animals by transiently occluding the aorta using a clip or forceps (Isbir et al. 2003).

Inflammation resulting from autoimmune disease can also damage the spinal cord. One animal model of autoimmune disease that targets the central nervous system and has been used to evaluate EPO as a potential therapeutic agent is experimental autoimmune encephalitis (EAE) (Agnello et al. 2002) . EAE can be induced in animals by immunization against myelin basic protein or myelin oligodendrocyte glycoprotein (Burkhardt et al. 1997). This procedure results in symptoms ranging from ataxia, hind limb paralysis and death. In experimental models, blunt trauma to the spinal cord may respond to a single dose of pharmacological treatment. In contrast, SCI resulting from autoimmune disease, a chronic condition, is likely to require multiple, ongoing treatments.

2.2 Primary and secondary mechanisms of injury

The initial damage to the spinal cord (such as mechanical deformation of the spine) may kill outright or only injure nerve cells, axons and blood vessels at the site of injury, leading to vasoconstriction, hemorrhage, and ischemia (Tator 1995). The primary injury, in turn, initiates a complex cascade of secondary injury including critical regional blood flow alterations, calcium-mediated cellular injury, fluid-electrolyte imbalance, free radical generation, glutamate-induced excitoxicity, disturbances in mitochondrion function, pro-inflammatory cytokine production and apoptotic cell death. The cellular debris from dead and dying cells subsequently attracts the attention of inflammatory cells, including circulating neutrophils and macrophages and resident microglia in the spinal cord. These cells further amplify injury by releasing pro-inflammatory cytokines at the site of injury and increase the damage to surrounding cells (Norenberg et al. 2004). Secondary mechanisms damage spinal cord tissue that was otherwise spared during the initial insult (Tator et al. 1991; Tator 1995) .

The evolutionary solution to the problem of the slow spread of cellular injury is the creation of a safe zone or "penumbra" by triggering cells surrounding the focal lesion to undergo programmed cell death (Erbayraktar et al. 2003). In SCI a penumbra of neurons at risk of undergoing programmed cell death surrounds the initial site of injury. A second wave of cell death targets oligodendrocytes, the cells that provide the myelin for axons within the spinal cord. Indeed, much of the motor and sensory paralysis that occurs after SCI is due to a delayed and widespread oligodendrocyte apoptosis and demyelination of long spinal tracts in the white matter (Beattie et al. 2002; Dong et al. 2003). The time delay elapses before oligodendrocyte cell death offers a window of opportunity for pharmacological intervention targeted at reducing the extent of secondary damage and improving clinical course. Notably, EPO plays a critical role in determining the size of the penumbra (see below).

As secondary injury subsides during the weeks and months following trauma, the site of spinal cord injury becomes typically characterized by disrupted axons and a cystic cavity encased within a glial scar with variable amounts of intact tissue surrounding the lesion. In this peripheral rim of intact tissue reside neurons that are either uninjured or that have lost part of their myelin sheaths. These neurons have the potential to regenerate axons and restore function. However, axonal regeneration after injury often fails because of elements within the injury environment that inhibit axonal growth and because the central nervous system neurons themselves demonstrate a relatively weak intrinsic ability to regenerate axons after injury (reviewed in Ref. 2).

3. ROLES FOR ENDOGENOUS EPO IN THE SPINAL CORD

EPO and its receptor, EPOR, have been shown to play important roles in the spinal cord in both normal and pathological conditions.

3.1 Normal EPO and EPOR expression in the spinal cord

Extensive work over the past few years has suggested a major developmental role for EPO in the nervous system. Many neurons and glial cells in the fetal spinal cord express EPO and EPOR (Juul et al. 1998; Juul et al. 1999). Although the levels of EPO and EPOR greatly diminish after birth, the spinal cord continues to express both molecules in a cell type specific manner consistent with a paracrine/autocrine role, with the ligand in close proximity to its target (Juul 2002; Buemi et al. 2003). Immunohistochemical staining of normal human spinal cord shows EPOR localized on capillaries, especially within the white matter (Celik et al. 2002; Sekiguchi et al. 2003). The somata and proximal dendrites of large motor neurons of the ventral horn are another site of EPOR expression, and are surrounded by a dense plexus of EPO-immunoreactive fibers (Celik et al. 2002).

3.2 Expression of EPO following SCI

Activation of hypoxia-inducing factor and/or pre-exposure to hypoxia protects neuronal cells from apoptosis induced, not only by hypoxia (Semenza 2001), but also by oxidative stress (Zaman et al. 1999). This response requires EPO, whose expression is part of the physiological response to hypoxia. EPO not only protects neuronal cells *in vitro* from apoptosis induced by hypoxia, but also from variety of other agents, including excitotoxins and glucose deprivation (reviewed in other chapters).

Within the penumbra surrounding ischemic brain lesions, cells express increased levels of EPOR and EPO. Temporally, EPOR upregulation occurs first, by ~12 hours, primarily in neurons and endothelial cells of the microcirculation, and is followed later by the increase in EPO expression by both astrocytes and neurons (Bernaudin et al. 1999). A recent study has shown that following the induction of EPO expression, this region is characterized by an increased resistance to subsequent stressors — a phenomenon known as delayed preconditioning (Dawson 2002).

If sufficient EPO reaches the EPOR-expressing cells in the penumbra, apoptosis is inhibited and the final region of injury is minimized. (EPO cannot save cells undergoing necrosis within the epicenter.) By limiting the extent of cell death EPO reduces the inflammatory response, thus reducing secondary injury (Brines et al. 2000; Ghezzi et al. (in press)). However, two factors act to reduce the probability that sufficient endogenous EPO reaches cells at risk before they become irreversibly committed to apoptosis. First the long latency for EPO production implies that EPO may be produced too late for optimal tissue protection. Second, and more importantly, proinflammatory cytokines directly inhibit EPO production (Nagai et al. 2001). Thus, although EPOR expressing cells are present in the penumbra, little EPO may be available and the cells die. Administration of exogenous recombinant human EPO circumvents these problems by assuring that adequate EPO is available within the penumbra.

4. EPO AS A NEUROPROTECTIVE AGENT IN SCI

Several investigations have suggested neuroprotective roles for EPO, but these have only recently been evaluated in spinal cord injury. To date, exogenously administered EPO has been shown to produce significant and prolonged neuroprotection in models of transient spinal cord ischemia (Celik et al. 2002), traumatic spinal cord injury (Gorio et al. 2002; Kaptanoglu et al. 2004), spinal nerve root crush injury (Sekiguchi et al. 2003), and spinal cord inflammation in EAE (Agnello et al. 2002).

4.1 Rabbit ischemia model

A transient spinal ischemia model in rabbits has been useful to evaluate EPO's effectiveness in ameliorating spinal cord ischemic injuries (Celik et al. 2002). The infrarenal aorta was exposed and non-traumatically occluded for a period of time, causing transient spinal cord ischemia. One of the attractive features of this model is the unique segmental arterial blood supply from the infrarenal aorta in the rabbit, which prevents bowel and kidney ischemia and thus avoids non-neurological complications.

Treating animals with recombinant human EPO immediately after the release of occlusion was associated with a significant improvement in neurologic score within one hour of treatment. In contrast, saline-treated animals did not significantly improve over the following 48 hours. In saline-treated animals the gray matter was grossly disrupted and infiltrated by abundant inflammatory cells; most neurons were shrunken and pyknotic or necrotic. In contrast, tissue sections obtained from rhEPO-treated animals often lacked histological evidence of injury. Furthermore, in saline-treated animals spinal cord sections adjacent to the site of injury showed many motor neurons positive for TUNEL labeling (a marker of apoptosis). Ischemic spinal cords from rhEPO-treated animals were generally devoid of such staining.

The study demonstrated that, in spite of its relatively large size of ~34 kD, systemically-administered rhEPO readily crosses the blood-spinal cord barrier. When given immediately after reperfusion, rhEPO improved the neurological damage following transient spinal ischemia, presumably by reducing apoptosis and inflammation.

4.2 Rodent contusion and compression models

The neuroprotective effects of rhEPO were also tested in compression and contusion models of SCI employing a modified aneurysm clip for one minute (moderately severe injury and relatively little hemorrhage) or by impaction with a steel rod (severe disruption of white and grey matter) (Gorio et al. 2002). In both cases, rhEPO was administered via an intraperitoneal injection immediately after injury.

Animals receiving rhEPO exhibited improvement of motor function at a time at which saline treated animals were still completely paralyzed and by the end of the study, at 28 days, treated animals had a markedly superior clinical course, with treated animals exhibiting a nearly normal score (18 out of 21) compared to placebo (score of 10). Animals in the compression model group showed improvement as early as 12 hours after treatment.

In the contusion model, characterized by more extensive damage, the treated group had a remarkable preservation of white matter, whereby myelinated axons appeared histologically normal. In contrast, animals receiving saline exhibited widespread degeneration and swollen myelin sheaths. In saline-treated animals apoptotic oligodendrocytes were detected by TUNEL labeling in the region of the *fasciculus cuneatus*. In contrast, no apoptotic nuclei were observed in this region in the rhEPO-treated animals. In addition, the treatment group exhibited an obvious reduction in the number of inflammatory cells in and around the region of injury.

Kaptanoglu and others (Kaptanoglu et al. 2004) confirmed the neuroprotective effects of rhEPO in another contusion model of SCI. In this study, a single intraperitoneal administration of EPO resulted in effective preservation of the spinal cord ultrastructure as determined by electron microscopic examination. The authors compared the ultrastructural neuroprotection afforded by rhEPO to that of methylprednisolone and found that the two treatments gave similar results. However, EPO was more effective than methylprednisolone in inhibiting SCI-induced lipid peroxidation, a process that leads to extensive cellular damage.

4.3 Nerve root crush injury model

To produce a rat model of nerve root injury, the main cause of lower back pain in humans, the L5 dorsal root ganglia was exposed and the nerve crushed with fine forceps for two seconds (Sekiguchi et al. 2003). Animals were treated with either vehicle or rhEPO one day before injury and daily after that. In this model 80% of ipsilateral hindpaws exhibited mechanical allodynia one and three days post nerve root crush, in both treatment and placebo groups. By day 7, 50% of ipsilateral hindpaws in untreated animals demonstrated allodynia, compared to only 12.5% of animals treated with rhEPO. In this model, although EPO was not initially protective, it significantly augmented the recovery rate from mechanical allodynia by one week. The effects of rhEPO were sustained throughout the experimental period.

4.4 Autoimmune disease

A recent study by Agnello and others (Agnello et al. 2002) examined the effect of systemic rhEPO administration on inflammation in the spinal cord in a rat model of EAE, where the disease was induced by immunization with myelin basic protein (MBP). EAE was associated with significant inflammation in the spinal cord. Administration of rhEPO daily from day 3 after immunization with MBP delayed the onset of EAE and decreased the clinical score. RhEPO-treated animals exhibited reduced inflammation and glial activation/proliferation in the spinal cord. The clinical effects were maintained without further EPO administration for at least 2 months.

5. MECHANISM OF NEUROPROTECTION

Although the exact mechanisms of EPO's neuroprotective effect are not fully understood, the promotion of cell survival signaling cascades; attenuation of intracellular calcium and nitric oxide production; restoration of vascular autoregulation; and antioxidative and anti-inflammatory processes have all been implicated as possible mechanisms (Erbayraktar et al. 2003). Other developmental roles of EPO may also come into play with important effects on restoring function by repair and regeneration of neuronal processes (Shingo et al. 2001).

5.1 Blocking apoptosis

In vitro, EPO protects neurons from cell death induced by hypoxia and a variety of other agents, including excitotoxins and glucose deprivation (Ghezzi et al. 2004). In contrast to its antiapoptotic effect, EPO has not shown any activity in preventing necrosis in the nervous system (Sinor et al. 2000). *In vivo*, several studies examining SCI in animal models have documented an effect of rhEPO on apoptosis. Following crush injury of a spinal nerve root EPO prevented apoptosis in dorsal root ganglion neurons (Sekiguchi et al. 2003), which was accompanied by an improvement in pain behavior. Similarly, in a transient spinal cord ischemia model, there was no detectable TUNEL labeling in ventral horn motor neurons (Celik et al. 2002), strongly supportive of an effect of EPO in blocking the induction of an apoptotic program in this class of neurons. A lack of inflammatory infiltrate, also noted by the authors, probably arises from inhibition of apoptosis and/or of a proximal member of the inflammatory cascade, such as caspase-1.

EPO is known to augment the number of circulating erythrocytes by preventing the apoptosis of late erythroid progenitors (Koury et al. 1990; Yu et al. 1993). Several antiapoptotic pathways regulated by EPO have been identified in erythroid precursors (Wojchowski et al. 1999) and are implicated in the prevention of neuronal apoptosis. EPO activates Janus kinase-2 (Jak2) and STAT5, which in turn induce survival proteins including Bcl-xL (Silva et al. 1999). In addition, EPO has been shown to activate phosphatidylinositol-3 kinase (PI3K) in the UT-7 leukemia cell line, where it recruits Akt (Kashii et al. 2000). This PI3K-Akt pathway also leads to upregulation of Bcl-xL (Leverrier et al. 1999). NF-kB is also a target of the PI3k-Atk pathway and mediates antiapoptotic signaling through platelet-derived growth factor (Romashkova et al. 1999). Thus, several pathways may be responsible for EPO's antiapoptotic effects.

5.2 Reducing inflammation

Inflammatory cells are involved in both the early and late damage that occurs in SCI. EPO appears to reduce the inflammatory infiltration at the injury site (Gorio et al. 2002). Methylprednisolone and ganglioside GM1, two agents that have been reported to be partially effective in the contusion and compression animal models of SCI, on the other hand, do not reduce the infiltration of neutrophils immediately after SCI (Taoka et al. 2000).

Using an *in vitro* system of neurons co-cultured with glia, Villa and others (Villa et al. 2003) have shown that the anti-inflammatory effects of EPO are not explained by a direct antagonism or a reduction of pro-inflammatory cytokines but rather by an antiapoptotic effect on neurons. Neurons that are exposed to trimethyltin (TMT), a toxin that induces neuronal apoptosis, release factors that stimulate tumor necrosis factor (TNF) production by glial cells which, in turn, amplify neurotoxicity. In this context EPO inhibits TNF production by TMT-exposed neuron-glia co-cultures. EPO does not, however, inhibit TNF production by pure glial preparations directly exposed to neuronal products or LPS.

In contradistinction, a direct anti-inflammatory effect of EPO was demonstrated in a model of EAE, a disease characterized by marked inflammation in the almost complete absence of neuronal death, at least in the earliest times of the disease process (Agnello et al. 2002). EPO does not affect autoimmunity in general, as it was not effective in a model of arthritis induced in rats.

One way EPO could affect inflammation is through modulation of members of the nuclear factor NF-kB, which is a major regulator of inflammatory genes. NF-kB has been shown to be strongly up-regulated after SCI by macrophages/microglia, endothelial cells, and neurons (Bethea et al. 1998). More recently, EPO has been shown to engage the NF-kB pathway, as well as the Jak2 pathway (Digicaylioglu et al. 2001). Thus, there appears to be cross-talk between the antiapoptotic and anti-inflammatory effects of EPO.

5.3 Restoring vascular integrity

One of the earliest changes following experimental traumatic SCI is a profound reduction in spinal cord blood flow that progressively worsens and may last for many hours. The very early recovery observed after EPO administration in several animal models of SCI could depend on a beneficial effect of EPO on the restoration of blood flow after injury.

Indeed EPO has been shown *in vitro* to antagonize the apoptosis of endothelial cells subjected to ischemic stressors (Chong et al. 2002). In addition, EPO has been shown to stimulate mitogenesis and support angiogenesis, both functioning to improve tissue oxygenation. EPO has also been shown to strengthen the tight junctions of endothelial cells and therefore reduce leakage from the capillaries within the brain parenchyma (Martinez-Estrada et al. 2003).

5.4 Other tissue-protective and tissue-regenerative effects

EPO appears to have multiple effects on neurons that may also impact the recovery from SCI. Activation of the EPOR is associated with a reduction of

Ca+ influx upon depolarization, thereby reducing synaptic vesicle release of potentially neurotoxic compounds, such as glutamate (Kawakami et al. 2001). EPO treatment increases astrocyte production of glutathione peroxidase and in this manner ameliorates neuronal damage caused by excitotoxins (Genc et al. 2002). Kaptanoglu and others (Kaptanoglu et al. 2004) have shown that EPO administered immediately after acute SCI in a rat model reduces lipid peroxidation by-products and protects spinal cord from injury.

In addition to limiting the extent of damage following SCI, EPO may also enhance neuronal regeneration (Shingo et al. 2001). Neural stem cells (NSCs) present in the brain and spinal cord proliferate to form spheres of undifferentiated cells that produce neurons, astrocytes, and oligodendrocytes, as well as precursors to secondary spheres (self-renewal). Shingo and others (Shingo et al. 2001) found that cultured NSCs exposed to EPO produce two- to three-fold more neurons, and this effect correlated with a reduced number of secondary multipotent NSCs. Thus, EPO might contribute to recovery after SCI by increasing the number of new neurons, helping to replace those lost to injury.

6. TRANSLATION TO HUMAN TRIALS

Recombinant human EPO has been widely used in clinical practice to treat anemia associated with renal failure, cancer and surgery and has an excellent safety profile. As animal models have demonstrated that single doses of rhEPO are remarkably effective for the treatment of acute SCI, there is good reason to be optimistic that rhEPO (or one of its analogues) may show similar results in clinical trials.

6.1 Therapeutic window

There is a wide therapeutic window before the damage following SCI becomes irreversible. In animal models treatment by rhEPO up to 24 hours after the initial injury is almost as effective as treatment given at the time of injury. The timing probably reflects the time that elapses before cells become committed to the apoptotic pathway. Unlike EPO, methylprednisolone therapy is ineffective and can worsen injury in rat SCI models (ref. 58; Brines et al., unpublished observations), whereas in human trials it has been reported to have some effect when administered within 8 hours, but worsens it when administered later (Erbayraktar et al. 2003).

If rhEPO is given immediately after acute injury, one dose is very effective for recovery of neurological function. However, if treatment is delayed, multiple doses may be more effective—a late treatment permits the initiation of multiple signaling pathways that lead to tissue damage and cell death at different time points.

6.2 Non-hematopoietic EPO derivatives

A principal concern with EPO therapy is that administration of several doses of this protein may lead to a potentially harmful increase in red cell mass and also for thrombosis. To address this concern, a number of derivatives have been developed that retain the tissue protective effects of EPO but not its hematopoietic activity.

One approach is based on the understanding that the effective production of red cells requires a continuous presence of EPO, whereas only brief exposure is sufficient for neuroprotection *in vitro*. Thus, a short-lived EPO could be translocated into tissue beds to initiate neuroprotection through EPOR activation but not survive long enough within the circulation to stimulate erythropoiesis. To produce a short-lived EPO the sialic acids that delay clearance *in vivo* were removed (Erbayraktar et al. 2003). In confirmation, systemically administered asialo-erythropoietin (asialoEPO) does not increase erythrocyte mass, yet is fully protective in animal models of stroke, spinal cord injury and peripheral neuropathy. Multiple doses of asialoEPO were as effective as rhEPO in restoring motor function after spinal cord compression using an aneurysm clip model (Erbayraktar et al. 2003).

Other classes of new EPO derivatives exploit the fact that hematopoietic and tissue-protective functions of EPO are mediated by interaction with different receptors. AsialoEPO binds the same receptor as EPO but with an extremely short plasma half-life. In contrast, carbamylated EPO (CEPO) and certain EPO mutants (Leist et al. 2004) do not bind at all to the classical EPOR receptor and therefore are without hematopoietic activity in human cell culture assays or upon chronic dosing in different animal models.

Notably, CEPO was significantly neuroprotective in a model of spinal cord compression where animals were treated for over 6 weeks (Leist et al. 2004). Even when the first dose was given with a delay of 48 hours or 72 hours after injury (Figure 1A), a significant beneficial effect on neurological function was observed, compared to the saline and EPO groups (Figure 1B). In addition, the neurological deficits in mice immunized with myelin oligodendrocyte glycoprotein to induce EAE were improved by CEPO over a prolonged observation period. Even 4 weeks after the robust plateau of neurological dysfunction was reached, a three times per week treatment with CEPO significantly improved neurological function.

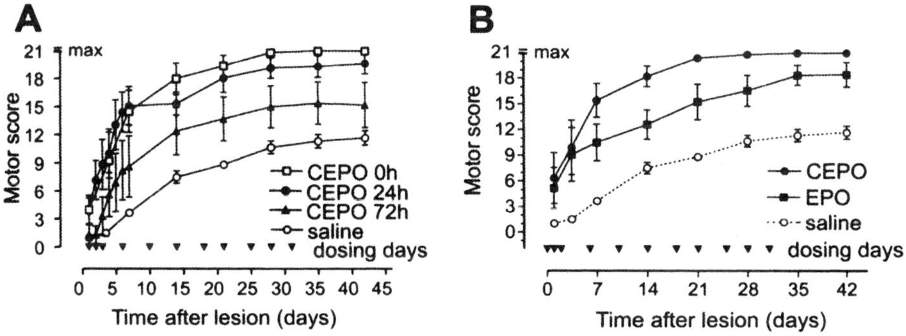

Figure 9-1. **A.** Temporal window of effectiveness of carbamylated EPO (CEPO) administered as a single dose (10 ug/kg iv) immediately following a 1 minute compression of rat spinal cord according to the methodology of Leist et al. (Leist et al. 2004) A motor score of 0 is complete paralysis and 21 is normal. CEPO provides substantial benefit compared to saline even when administered 3 days following injury. Tissue protective cytokines show some efficacy in this model when dosing begins less than 7 days following injury but none thereafter (Brines et al., unpublished observations) **B.** Comparison of single doses of CEPO or EPO administered immediately following injury in a rat model of SCI performed as in Figure 1 shows that CEPO is superior to EPO. (Reproduced with permission of *Science*.)

7. CONCLUSION

There are currently few options available to physicians treating patients who suffer from SCI. One possible strategy is to develop agents to block the cascade of secondary damage pathways that follow the initial injury to the cord. So far, four pharmacological substances have met rigorous criteria in laboratory testing and initial human investigations: two corticosteroids (methylprednisolone and tirilazad mesylate), naloxone, and GM-1 ganglioside (American Association of Neurological Surgeons 2002). The available medical evidence does not support a significant clinical benefit for any of these substances. While a neurological recovery benefit of methylprednisolone when administered within 8 hours of SCI has been suggested in some studies, recent meta analyses suggest that no such benefit exists. Importantly, the administration of methylprednisolone for 24 hours has been clearly associated with a significant increase in severe medical complications (American Association of Neurological Surgeons 2002).

Unlike these compounds, and others being evaluated in SCI therapy, rhEPO and its non-erythropoietic analogues target an endogenous tissue-protective system that involves many levels of activity. EPO is a principal modifier of apoptosis in multiple cell types in different tissues and organs in the setting of potential injury. However, EPO also exhibits many other actions that serve to protect cells either directly or indirectly. It can thus act at many time points in the cascade of spinal cord injury to block damaging pathways (Ghezzi et al. (in press)). Recent evidence that EPO may also contribute to the regeneration of neurons following injury warrants further investigation (Shingo et al. 2001).

Millions of patients have received rhEPO for the treatment of anemia, and it has been shown in general to be an exceedingly safe drug. The recent phase II trial of human stroke by rhEPO (Ehrenreich et al. 2002) used three daily doses substantially higher than the clinical norm and in these patients rhEPO was shown to be safe and did not raise the hemoglobin concentration above normal. Thus rhEPO might represent a valuable therapeutic approach to acute CNS injury, such as spinal cord trauma. A number of rhEPO analogs that retain rhEPO's neuroprotective functions but not its hematopoietic effects have been developed (Erbayraktar et al. 2003; Leist et al. 2004) and will likely broaden the range of possible therapeutic application, especially for conditions requiring multiple doses.

Clinical trials to assess the efficacy of therapy of rhEPO or one of its analogs should initially focus on isolated spinal cord injury, as the extremely heterogeneous nature of associated head and widespread organ damage will likely complicate clinical trials. Clinical evaluation end points should include motor, sensory and autonomic function and include a quality of life assessment.

Because of the devastating consequences of SCI, effective treatment is urgently needed. The outstanding safety record of rhEPO treatment in anemia and the demonstration of rhEPO and its derivatives' broad neuroprotective effects in animal models should encourage early evaluation of tissue protective cytokines in the setting of SCI.

ACKNOWLEDGEMENTS

The initial impetus to study spinal cord injury was provided by our colleagues Necati Gokman, Serhat Erbayraktar and Osman Yilmaz at Dokuz Eylul University Medical School, Izmir, Turkey.

REFERENCES

Agnello, D., Bigini, P., Villa, P., Mennini, T., Cerami, A., Brines, M. L. and Ghezzi, P., 2002, Erythropoietin exerts an anti-inflammatory effect on the CNS in a model of experimental autoimmune encephalomyelitis, *Brain Res* **952**: 128-34.

American Association of Neurological Surgeons, 2002, Pharmacological therapy after acute cervical spinal cord injury, *Neurosurgery Suppl.* **50**: S63-S72.

Beattie, M. S., Hermann, G. E., Rogers, R. C. and Bresnahan, J. C., 2002, Cell death in models of spinal cord injury, Prog Brain Res **137**: 37-47.

Bernaudin, M., Marti, H. H., Roussel, S., Divoux, D., Nouvelot, A., MacKenzie, E. T. and Petit, E., 1999, A potential role for erythropoietin in focal permanent cerebral ischemia in mice, J Cereb Blood Flow Metab **19**: 643-51.

Bethea, J. R., Castro, M., Keane, R. W., Lee, T. T., Dietrich, W. D. and Yezierski, R. P., 1998, Traumatic spinal cord injury induces nuclear factor-kappaB activation, J Neurosci **18**: 3251-60.

Blight, A. R. and Zimber, M. P., 2001, Acute spinal cord injury: pharmacotherapy and drug development perspectives, Curr Opin Investig Drugs **2**: 801-8.

Brines, M. L., Ghezzi, P., Keenan, S., Agnello, D., de Lanerolle, N. C., Cerami, C., Itri, L. M. and Cerami, A., 2000, Erythropoietin crosses the blood-brain barrier to protect against experimental brain injury, Proc Natl Acad Sci U S A **97**: 10526-31.

Buemi, M., Cavallaro, E., Floccari, F., Sturiale, A., Aloisi, C., Trimarchi, M., Corica, F. and Frisina, N., 2003, The pleiotropic effects of erythropoietin in the central nervous system, J Neuropathol Exp Neurol **62**: 228-36.

Burkhardt, H. and Kalden, J. R., 1997, Animal models of autoimmune diseases, Rheumatol Int **17**: 91-9.

Carlson, G. D. and Gorden, C., 2002, Current developments in spinal cord injury research, Spine J **2**: 116-28.

Celik, M., Gokmen, N., Erbayraktar, S., Akhisaroglu, M., Konakc, S., Ulukus, C., Genc, S., Genc, K., Sagiroglu, E., Cerami, A. and Brines, M., 2002, Erythropoietin prevents motor neuron apoptosis and neurologic disability in experimental spinal cord ischemic injury, Proc Natl Acad Sci U S A **99**: 2258-63.

Chong, Z. Z., Kang, J. Q. and Maiese, K., 2002, Erythropoietin is a novel vascular protectant through activation of Akt1 and mitochondrial modulation of cysteine proteases, Circulation **106**: 2973-9.

Dawson, T. M., 2002, Preconditioning-mediated neuroprotection through erythropoietin?, Lancet **359**: 96-7.

Digicaylioglu, M. and Lipton, S. A., 2001, Erythropoietin-mediated neuroprotection involves cross-talk between Jak2 and NF-kappaB signalling cascades, Nature **412**: 641-7.

Dong, H., Fazzaro, A., Xiang, C., Korsmeyer, S. J., Jacquin, M. F. and McDonald, J. W., 2003, Enhanced oligodendrocyte survival after spinal cord injury in Bax-deficient mice and mice with delayed Wallerian degeneration, J Neurosci **23**: 8682-91.

Dumont, R. J., Okonkwo, D. O., Verma, S., Hurlbert, R. J., Boulos, P. T., Ellegala, D. B. and Dumont, A. S., 2001, Acute spinal cord injury, part I: pathophysiologic mechanisms, Clin Neuropharmacol **24**: 254-64.

Ehrenreich, H., Hasselblatt, M., Dembowski, C., Cepek, L., Lewczuk, P., Stiefel, M., Rustenbeck, H. H., Breiter, N., Jacob, S., Knerlich, F., Bohn, M., Poser, W., Ruther, E., Kochen, M., Gefeller, O., Gleiter, C., Wessel, T. C., De Ryck, M., Itri, L., Prange, H., Cerami, A., Brines, M. and Siren, A. L., 2002, Erythropoietin therapy for acute stroke is both safe and beneficial, Mol Med **8**: 495-505.

Erbayraktar, S., Grasso, G., Sfacteria, A., Xie, Q. W., Coleman, T., Kreilgaard, M., Torup, L., Sager, T., Erbayraktar, Z., Gokmen, N., Yilmaz, O., Ghezzi, P., Villa, P., Fratelli, M., Casagrande, S., Leist, M., Helboe, L., Gerwein, J., Christensen, S., Geist, M. A., Pedersen, L. O., Cerami-Hand, C., Wuerth, J. P., Cerami, A. and Brines, M., 2003, Asialoerythropoietin is a nonerythropoietic cytokine with broad neuroprotective activity in vivo, Proc Natl Acad Sci U S A **100**: 6741-6.

Erbayraktar, S., Yilmaz, O., Gokmen, N. and Brines, M., 2003, Erythropoietin is a multifunctional tissue-protective cytokine, Curr Hematol Rep **2**: 465-70.

Fisher, J. W., 2003, Erythropoietin: physiology and pharmacology update, *Exp Biol Med (Maywood)* **228**: 1-14.

Genc, S., Akhisaroglu, M., Kuralay, F. and Genc, K., 2002, Erythropoietin restores glutathione peroxidase activity in 1-methyl-4-phenyl-1,2,5,6-tetrahydropyridine-induced neurotoxicity in C57BL mice and stimulates murine astroglial glutathione peroxidase production in vitro, *Neurosci Lett* **321**: 73-6.

Gharagozloo, F., Larson, J., Dausmann, M. J., Neville, R. F., Jr. and Gomes, M. N., 1996, Spinal cord protection during surgical procedures on the descending thoracic and thoracoabdominal aorta: review of current techniques, *Chest* **109**: 799-809.

Ghezzi, P. and Brines, M., 2004, Erythropoietin as an antiapoptotic, tissue-protective cytokine, *Cell Death Differ* **11 (Suppl 1)**: S37-44

Gorio, A., Gokmen, N., Erbayraktar, S., Yilmaz, O., Madaschi, L., Cichetti, C., Di Giulio, A. M., Vardar, E., Cerami, A. and Brines, M., 2002, Recombinant human erythropoietin counteracts secondary injury and markedly enhances neurological recovery from experimental spinal cord trauma, *Proc Natl Acad Sci U S A* **99**: 9450-5.

Isbir, C. S., Ak, K., Kurtkaya, O., Zeybek, U., Akgun, S., Scheitauer, B. W., Sav, A. and Cobanoglu, A., 2003, Ischemic preconditioning and nicotinamide in spinal cord protection in an experimental model of transient aortic occlusion, *Eur J Cardiothorac Surg* **23**: 1028-33.

Juul, S., 2002, Erythropoietin in the central nervous system, and its use to prevent hypoxic-ischemic brain damage, *Acta Paediatr Suppl* **91**: 36-42.

Juul, S. E., Yachnis, A. T. and Christensen, R. D., 1998, Tissue distribution of erythropoietin and erythropoietin receptor in the developing human fetus, *Early Hum Dev* **52**: 235-49.

Juul, S. E., Yachnis, A. T., Rojiani, A. M. and Christensen, R. D., 1999, Immunohistochemical localization of erythropoietin and its receptor in the developing human brain, *Pediatr Dev Pathol* **2**: 148-58.

Kaptanoglu, E., Solaroglu, I., Okutan, O., Surucu, H. S., Akbiyik, F. and Beskonakli, E., 2004, Erythropoietin exerts neuroprotection after acute spinal cord injury in rats: effect on lipid peroxidation and early ultrastructural findings, *Neurosurg Rev* **27**: 113-20.

Kashii, Y., Uchida, M., Kirito, K., Tanaka, M., Nishijima, K., Toshima, M., Ando, T., Koizumi, K., Endoh, T., Sawada, K., Momoi, M., Miura, Y., Ozawa, K. and Komatsu, N., 2000, A member of Forkhead family transcription factor, FKHRL1, is one of the downstream molecules of phosphatidylinositol 3-kinase-Akt activation pathway in erythropoietin signal transduction, *Blood* **96**: 941-9.

Kawakami, M., Sekiguchi, M., Sato, K., Kozaki, S. and Takahashi, M., 2001, Erythropoietin receptor-mediated inhibition of exocytotic glutamate release confers neuroprotection during chemical ischemia, *J Biol Chem* **276**: 39469-75.

Kawakami, M., Weinstein, J. N., Spratt, K. F., Chatani, K., Traub, R. J., Meller, S. T. and Gebhart, G. F., 1994, Experimental lumbar radiculopathy. Immunohistochemical and quantitative demonstrations of pain induced by lumbar nerve root irritation of the rat, *Spine* **19**: 1780-94.

Koury, M. J. and Bondurant, M. C., 1990, Erythropoietin retards DNA breakdown and prevents programmed death in erythroid progenitor cells, *Science* **248**: 378-81.

Kwon, B. K., Borisoff, J. F. and Tetzlaff, W., 2002, Molecular targets for therapeutic intervention after spinal cord injury, *Mol Intervent* **2**: 244-58.

Leist, M., Ghezzi, P., Grasso, G., Bianchi, R., Villa, P., Fratelli, M., Savino, C., Bianchi, M., Nielsen, M., Torup, L., Sager, T., Sfacteria, A., Erbayraktar, S., Erbayraktar, Z., Gokmen, N., Yilmaz, O., Cerami-Hand, C., Xie, Q. W., Coleman, T., Cerami, A. and Brines, M., 2004, Erythropoietin-derived tissue-protective cytokines that do not bind to the classical erythropoietin receptor., *Science* **305**: 239-242.

Leverrier, Y., Thomas, J., Mathieu, A. L., Low, W., Blanquier, B. and Marvel, J., 1999, Role of PI3-kinase in Bcl-X induction and apoptosis inhibition mediated by IL-3 or IGF-1 in Baf-3 cells, *Cell Death Differ* **6**: 290-6.

Martinez-Estrada, O. M., Rodriguez-Millan, E., Gonzalez-De Vicente, E., Reina, M., Vilaro, S. and Fabre, M., 2003, Erythropoietin protects the in vitro blood-brain barrier against VEGF-induced permeability, *Eur J Neurosci* **18**: 2538-44.

Nagai, A., Nakagawa, E., Choi, H. B., Hatori, K., Kobayashi, S. and Kim, S. U., 2001, Erythropoietin and erythropoietin receptors in human CNS neurons, astrocytes, microglia, and oligodendrocytes grown in culture, *J Neuropathol Exp Neurol* **60**: 386-92.

National Spinal Cord Injury Statistical Center, 2003, http://www.spinalcord.uab.edu.

Norenberg, M. D., Smith, J. and Marcillo, A., 2004, The pathology of human spinal cord injury: defining the problems, *J Neurotrauma* **21**: 429-40.

Rabchevsky, A. G., Fugaccia, I., Sullivan, P. G., Blades, D. A. and Scheff, S. W., 2002, Efficacy of methylprednisolone therapy for the injured rat spinal cord, *J Neurosci Res* **68**: 7-18.

Romashkova, J. A. and Makarov, S. S., 1999, NF-kappaB is a target of AKT in anti-apoptotic PDGF signalling, *Nature* **401**: 86-90.

Rosenzweig, E. S. and McDonald, J. W., 2004, Rodent models for treatment of spinal cord injury: research trends and progress toward useful repair, *Curr Opin Neurol* **17**: 121-31.

Sakanaka, M., Wen, T. C., Matsuda, S., Masuda, S., Morishita, E., Nagao, M. and Sasaki, R., 1998, In vivo evidence that erythropoietin protects neurons from ischemic damage, *Proc Natl Acad Sci U S A* **95**: 4635-40.

Sasaki, R., 2003, Pleiotropic functions of erythropoietin, *Intern Med* **42**: 142-9.

Sekiguchi, Y., Kikuchi, S., Myers, R. R. and Campana, W. M., 2003, ISSLS prize winner: Erythropoietin inhibits spinal neuronal apoptosis and pain following nerve root crush, *Spine* **28**: 2577-84.

Semenza, G. L., 2001, HIF-1 and mechanisms of hypoxia sensing, *Curr Opin Cell Biol* **13**: 167-71.

Shingo, T., Sorokan, S. T., Shimazaki, T. and Weiss, S., 2001, Erythropoietin regulates the in vitro and in vivo production of neuronal progenitors by mammalian forebrain neural stem cells, *J Neurosci* **21**: 9733-43.

Short, D. J., El Masry, W. S. and Jones, P. W., 2000, High dose methylprednisolone in the management of acute spinal cord injury - a systematic review from a clinical perspective, *Spinal Cord* **38**: 273-86.

Silva, M., Benito, A., Sanz, C., Prosper, F., Ekhterae, D., Nunez, G. and Fernandez-Luna, J. L., 1999, Erythropoietin can induce the expression of bcl-x(L) through Stat5 in erythropoietin-dependent progenitor cell lines, *J Biol Chem* **274**: 22165-9.

Sinor, A. D. and Greenberg, D. A., 2000, Erythropoietin protects cultured cortical neurons, but not astroglia, from hypoxia and AMPA toxicity, *Neurosci Lett* **290**: 213-5.

Siren, A. L., Fratelli, M., Brines, M., Goemans, C., Casagrande, S., Lewczuk, P., Keenan, S., Gleiter, C., Pasquali, C., Capobianco, A., Mennini, T., Heumann, R., Cerami, A., Ehrenreich, H. and Ghezzi, P., 2001, Erythropoietin prevents neuronal apoptosis after cerebral ischemia and metabolic stress, *Proc Natl Acad Sci U S A* **98**: 4044-9.

Taoka, Y. and Okajima, K., 2000, Role of leukocytes in spinal cord injury in rats, *J Neurotrauma* **17**: 219-29.

Tator, C. H., 1995, Update on the pathophysiology and pathology of acute spinal cord injury, *Brain Pathol* **5**: 407-13.

Tator, C. H. and Fehlings, M. G., 1991, Review of the secondary injury theory of acute spinal cord trauma with emphasis on vascular mechanisms, *J Neurosurg* **75**: 15-26.

Villa, P., Bigini, P., Mennini, T., Agnello, D., Laragione, T., Cagnotto, A., Viviani, B., Marinovich, M., Cerami, A., Coleman, T. R., Brines, M. and Ghezzi, P., 2003, Erythropoietin selectively attenuates cytokine production and inflammation in cerebral ischemia by targeting neuronal apoptosis, *J Exp Med* **198**: 971-5.

Wojchowski, D. M., Gregory, R. C., Miller, C. P., Pandit, A. K. and Pircher, T. J., 1999, Signal transduction in the erythropoietin receptor system, *Exp Cell Res* **253**: 143-56.

Yu, H., Bauer, B., Lipke, G. K., Phillips, R. L. and Van Zant, G., 1993, Apoptosis and hematopoiesis in murine fetal liver, *Blood* **81**: 373-84.

Zaman, K., Ryu, H., Hall, D., O'Donovan, K., Lin, K. I., Miller, M. P., Marquis, J. C., Baraban, J. M., Semenza, G. L. and Ratan, R. R., 1999, Protection from oxidative stress-induced apoptosis in cortical neuronal cultures by iron chelators is associated with enhanced DNA binding of hypoxia-inducible factor-1 and ATF-1/CREB and increased expression of glycolytic enzymes, p21(waf1/cip1), and erythropoietin, *J Neurosci* **19**: 9821-30.

Chapter 10

ERYTHROPOIETIN AND NEUROPROTECTION IN THE PERIPHERAL NERVOUS SYSTEM: *IN VIVO* STUDIES

W. Marie Campana, Ph.D.
Department of Anesthesiology, University of California, San Diego, La Jolla, CA 92093-0629

Abstract: Erythropoietin (EPO) is now known to have pleiotrophic effects outside the hematological system. Systemically delivered recombinant human EPO (rhEPO) can cross the blood brain barrier and inhibit neuronal death and inflammatory infiltrates induced by ischemia in the brain. Recent evidence also suggests that rhEPO is a potent neuroprotective agent in primary sensory neurons and peripheral glia involved in pain transmission. When L5 spinal nerve is crushed, apoptosis of sensory neurons and satellite cells occurs and is inhibited by systemic treatment with rhEPO. In addition, the neuroprotective effects of rhEPO are observed in second order neurons of the spinal dorsal horn after nerve root. In both animal models, prevention of apoptosis in neurons by rhEPO in either peripheral or central locations correlates with facilitated recoveries from neuropathic pain states. However, the *in vivo* effects of rhEPO are not limited to direct effects on neurons. System administration of rhEPO in animals given a chronic constriction injury to the sciatic nerve, a model known to produce the pro-inflammatory cytokine, TNF-α, at the injury site results in significant reductions in Schwann cell produced TNF-α and faster recovery from neuropathic pain states. Thus, EPO may reduce glial activation associated with axonal degeneration and pain. In rodent models of diabetes, rhEPO ameliorates both mechanical and thermal allodynia and prevents loss of epidermal unmyelinated C-fibers, suggesting a therapeutic potential for rhEPO in treating both the degenerative and painful manifestations of diabetic neuropathy. Novel EPO mimetics, including neuroactive peptides derived from prosaposin and carbamylated EPO appear to have neuroactivity without hematological side effects associated with chronic rhEPO therapy and are currently under investigation as a novel approach to treating painful peripheral neuropathies.

Key words: neuropathic pain, neuroprotection, axonal degeneration, Schwann cells, and EPO mimetics

1. INTRODUCTION

The development of novel therapeutics to protect neurons following damage is a key strategy towards the treatment of injured nerves and neurodegenerative diseases. Damage to nerves can induce apoptosis of their cell bodies, leading to loss of normal function. One aspect that is potentially associated with nerve damage is the development of chronic neuropathic pain. While the link between nerve damage leading to apoptosis and chronic pain is controversial, there are several reports that indicate neuronal death as an underlying mechanism in neuropathic pain states. Several neuropathic pain models, including chronic constriction injury (CCI) of sciatic nerve and L5 spinal nerve ligation (L5 SNL), have shown evidence of apoptosis in primary sensory neurons and spinal neurons that preceded or corresponded with peak periods of pain behavior (Whiteside and Munglani, 2001; Campana and Myers, 2003). It has also been reported that neurons in the spinothalamic tract undergo delayed apoptosis after spinal cord injury, which contributed to chronic central pain (Qiu et al., 2001). One of the proposed mechanisms underlying the connection between apoptosis and pain is attenuating inhibitory transmitter release due to death of inhibitory (i.e. GABAergic, glycinergic) spinal neurons. Moore et al. (2002) demonstrated that after partial nerve injury, loss of post-synaptic GABA-induced inhibition of excitatory input contributed to increased spinal neuron death. Collectively, these findings suggest an eventual relationship between chronic pain and cell death.

Currently, pharmacological therapies used to treat neuropathic pain, which includes opioids, anti-convulsants, nonsteroidal anti-inflammatory and local anesthetic agents, are generally ineffective or have substantial drawbacks due to side effects. Generally, these treatments do not protect neurons from dying and there have even been reports that chronic morphine treatment leading to "morphine tolerance" may be related to morphine-induced death of neurons (Mao et al., 2002). Thus, developing therapeutics that has both analgesic and repairing properties would be rewarding. Neurotrophic factors, such as nerve growth factor-beta (NGF) and neurotrophin-4 (NT-4), possess potent neuroprotective activity (Meakin and Shooter, 1992; Deckwerth et al., 1993), however, these factors are limited as therapeutics because of their tendency to induce hyperalgesia (Shu and Mendell, 1999). This chapter focuses on the recent discovery of erythropoietin (EPO) as a potent neuroprotective factor and its role in treating neuropathic pain. Originally, EPO was best characterized for its activity as a hematopoietic growth factor and is FDA-approved for the treatment of anemia during chronic renal failure. The *in vivo* studies reviewed here have led to the discovery that EPO may be highly active in protecting neurons and facilitating recovery from neuropathic pain states after nerve injury.

2. EPO AND THE PERIPHERAL NERVOUS SYSTEM

EPO was first described as a specific kidney-produced hematopoietic growth factor, critical to the proliferation of red blood cell precursors. However, over the last decade EPO has been shown to have pleiotrophic effects. Others and we have identified a role for EPO in the nervous system (Konishi et al., 1993; Digicaylioglu et al., 1995; Campana et al., 1998; Sakanaka et al., 1998). One of EPO's most potent effects is protecting neurons from death induced by ischemia (Siren et al., 2000), proinflammatory cytokines (Digicaylioglu and Lipton, 2001) and glutamate excitotoxicity (Kawakami et al., 2001). The presence of EPO and its receptor (EPOR) were identified in sciatic nerve (Campana and Myers, 2001) and were temporally regulated after injury. In uninjured nerve, EPO immunoreactivity was present in both axons and Schwann cells, but in low abundance. EPOR was also present in both axons and Schwann cells. After chronic constriction injury (CCI), EPO immunoreactivity was dramatically increased and consistent with this early (1 and 3 days after injury) increase in EPO protein, real-time quantitative polymerase chain reaction (qPCR) demonstrated a 10-fold increase in EPO mRNA (Li et al., 2005). These results are similar to hypoxia-induced increases in EPO production observed in brain after ischemic injury (Bernaudin et al., 2000). Interestingly, EPOR immunoreactivity and mRNA remained relatively unchanged early after nerve injury; however, EPOR mRNA increased significantly by day 7 post injury (Li et al., 2005). Similar results were observed with EPOR in retinal ganglion cells; after optic nerve axotomy, EPOR levels remained relatively constant (Weishaupt et al., 2004). Recently, we have begun to examine the role of endogenously produced EPO and EPOR in Schwann cells after ischemic injury and its potential role in nerve regeneration.

The impact of exogenously administered recombinant human EPO (rhEPO) on peripheral nerve, DRG and spinal cord during nerve injury has been investigated (Campana and Myers, 2001, 2003; Celik et al., 2002; Sekiguchi et al., 2003). These studies will be discussed below. We hypothesize that EPO is a neuroprotective agent for primary sensory neurons, spinal neurons and their supporting glial cells, and this activity enhances recovery from neuropathic pain states.

3. EFFECTS OF EPO IN MECHANICAL OR COMPRESSION NERVE INJURY

3.1 Sciatic nerve

Peripheral nerve injury can be induced by several different methods (ligatures, crush, partial nerve injury) in experimental animals and are well-characterized models of chronic neuropathic pain. We used chronic constriction injury (CCI), as it has both an ischemic and inflammatory component (Myers et al., 1993). In this model, the sciatic nerve is exposed unilaterally at the mid-thigh level and four 4.0 chromic gut ligatures are placed around the nerve with 1 mm spacing. The ligatures are tied until they just slightly constrict the diameter of the nerve (Bennett and Xie, 1988) and a short twitch is observed in the respective hind limb. In this model, both TNF-α mRNA and protein are upregulated in Schwann cells at the injury site (Wagner and Myers, 1996). Our hypothesis was that EPO alleviates neuropathic pain induced by CCI and that TNF-α antagonism is a critical mechanism underlying the therapeutic effects of EPO. To determine whether exogenous rhEPO had an effect on pain behaviors, we used two behavioral endpoints: 1) thermal nociceptive test (Hargraeves et al., 1988) and 2) von Frey testing for mechanical allodynia (Chaplan et al., 1994). Responses to thermal stimulation were obtained by measuring latency of hind paw withdrawal from a focal radiant heat source. This methodology is considered to measure changes to C-fiber mediated activity. Responses to mechanical stimulation were assessed by hind foot withdrawal from weighted filaments. This methodology is considered to measure changes in Aβ-fiber mediated activity. Initially after CCI, thermal responses demonstrated a profound decrease in latency of CCI paws from both rhEPO- and vehicle-treated animals, indicative of thermal hyperalgesia. However, although EPO-treated animals were not initially protected from thermal hyperalgesia, EPO-treated animals recovered much more quickly than vehicle-treated animals (Campana et al., 2004). In addition, the ameliorative effects of rhEPO were specific to injured nerve, as rhEPO-treated animals in the uninjured sham group demonstrated no analgesic effects. These data indicate that while EPO is not initially protective or acutely analgesic against pain states induced by traumatic nerve injury, it may stabilize or prevent long term changes in injured nerve fibers that contribute to facilitated recovery from chronic pain. The effects of rhEPO were not limited to changes in thermal hyperalgesia; rhEPO was also effective in reducing mechanical allodynia. Vehicle- and rhEPO- treated CCI animals initially showed a significant decrease (indicative of pain) in mechanical withdrawal thresholds, however, only in the vehicle treated group were low thresholds sustained for 14 days. Animals receiving rhEPO demonstrated a trend for greater thresholds (toward normal) than vehicle-treated animals, which reached significance by day 7 and maintained for the duration of the experiment. Collectively, these data suggest that animals treated with exogenous rhEPO had faster recoveries from both hyperalgesic and allodynic pain states after peripheral nerve injury.

The mechanism underlying rhEPO-induced changes in chronic pain behaviors after CCI are unclear. One plausible hypothesis is that EPO regulates pro-inflammatory cytokine levels or activity in the local milieu of injured nerve. TNF-α, a pro-inflammatory cytokine that orchestrates cytokine activity after nerve injury, spontaneously induces C-fiber firing (Sorkin et al., 1999), and has a direct causal relationship with neuropathic pain, was a likely candidate. EPO and TNF-α have been shown to regulate each other in the hematopoietic system: EPO regulates TNF-α levels (Braczkowski et al., 2001) and erythropoiesis is inhibited by TNF-α (Cooper et al., 2003). In the peripheral nerve, real-time qPCR showed a 15-fold increase in TNF-α mRNA on day 1, and 9-fold increases in TNF-α mRNA on day 5 (Campana et al., 2004), respectively, after CCI. Exogenous addition of rhEPO (2680 units/kg) resulted in a 50% reduction in TNF-α mRNA at the injury site on both these days. In addition, local endoneurial injection of rhEPO (400 pg) into nerve at the injury site, significantly reduced TNF-α mRNA levels. Thus, it appears that rhEPO regulates TNF-α transcriptionally; however, whether rhEPO stabilizes TNF-α mRNA or directly enhances transcriptional activity is not known. Interestingly, dual label immunofluorescence and confocal microscopy indicated that changes in TNF-α levels were consistently observed in Schwann cells. Co-localization of S100 and TNF-α positive cells was less frequently observed at the injury site after rhEPO treatment compared to vehicle-treated controls. We concluded that one of the mechanisms by which rhEPO modulates pain behavior and perhaps neuroprotection after injury is a consequence of TNF-α downregulation. Thus, "good" cytokines more abundantly located at the injury site than "bad" cytokines may help facilitate recovery from neuropathic pain states. Our findings are consistent with those observed in the central nervous system; rhEPO treatment downregulated TNF-α levels in animal models of multiple sclerosis (Villa et al., 2003), where TNF-α has deleterious effects.

3.2 Spinal Nerve

The L5 spinal nerve encompasses the end of the sciatic nerve and branches into the L5 DRG. In models of L5 spinal nerve crush (L5 SNC), the nerve is crushed 2 mm distal to the DRG for two seconds with flat forceps. This induces a sustained (>60 days) mechanical allodynia in rats (Winkelstein et al., 2001; Campana and Myers, 2003). In addition, the close proximity of nerve injury to the DRG induces chromatolysis in the cell body of the sensory neuron (Degn et al., 1999) and for more susceptible neurons, induces death (Campana and Myers, 2001). Because rhEPO has been shown to be a potent neuroprotective agent in the CNS (Siren et al., 2001), we tested whether systemic treatment with rhEPO protected primary sensory neurons from death after L5 SNC. First, we confirmed the presence of EPOR in the cell body of DRG neurons. We noted that EPOR was located in both satellite cells and cell bodies (Campana and Myers, 2003). The subcellular distribution of EPOR was shown to be punctate at the borders of the cell body with diffuse staining in the cytoplasm. In the satellite cells, perinuclear co-localization was apparent. Thus, systemic rhEPO could have direct effects on primary sensory neurons and their supporting glia, particularly since the blood-brain barrier of the DRG tends to be more permeable than that of the spinal cord.

The effective dose of systemically delivered rhEPO was determined by assessing pain behavior (mechanical allodynia). Systemic rhEPO was given subcutaneously at 1000 units/kg, 2680 units/kg, and 5000 units/kg doses. We incorporated a boiled dose at 2680 units/kg to ensure that denatured rhEPO protein was inactive. Animals were given rhEPO one day prior to L5 SNC and, subsequently, daily injections of rhEPO were given until day 14. Both the 2680 units/kg and 5000 units/kg rhEPO dosed animals showed significant increases in thresholds (towards normal) (Campana and Myers, 2003). These results corresponded with the prevention of apoptosis of both primary sensory neurons and satellite cells in rhEPO treated animals. Interestingly, the effects of rhEPO on apoptosis occurred very early on (1-2 days post-injury), whereas, the alleviation from mechanical allodynia occurred 1 week after L5 SNC. This suggests that early mechanisms (aberrant electrical discharge, changes in sodium channels, etc.) involved in nerve injury leading to neuropathic pain states are independent of rhEPO neuroprotective effects. However, rhEPO may facilitate a faster recovery from long-term neuropathic pain behaviors by protecting neurons and glia from death. The presence of viable cells would support faster regeneration and recovery from altered pain states.

EPOR signals through JAK2 in hematopoietic cells (Bittorf et al., 1994). We demonstrated that phosphorylated JAK2 was upregulated in DRG within one day after rhEPO treatment during L5 SNC. This preceded the anti-apoptotic effects observed in both neuronal cell bodies and satellite cells 2 days post L5 SNC. While the JAK2 phosphorylation and anti-apoptosis data remains correlative in primary sensory neurons, it has been previously shown that activation of JAK2 by EPO directly initiated a neuroprotective signaling pathway in cortical neurons (Digicaylioglu and Lipton, 2001). This was shown be using both a dominant negative construct to JAK2 and a pharmacological inhibitor, AG490. In DRG, phosphorylated JAK2 was localized to NF200-positive neurons (large, mechanosensitive neurons) and peptide-positive neurons (small, nociceptive neurons) and GFAP-positive glia (satellite cells). Thus, it is not surprising that systemic delivery of rhEPO prevented apoptosis in both neurons and glia in DRG.

3.3 Nerve Root

The protection of spinal cord neurons following injury to nerve roots is of primary importance in avoiding permanent functional deficits and protracted pain states. The nerve roots are connected to the spinal cord and have special relevance in terms of back pain and spinal anatomy. Nerve roots are ensheathed by arachnoid and dura mater and, therefore, are surrounded by cerebrospinal fluid. Scar formation secondary to nerve root injury and/or spinal surgery can cause pain. In addition, cytokines liberated by tissue injury or by herniated nucleus pulposes of spinal discs includes the primary inflammatory cytokine, TNF-α. Because changes in motor function and severe pain are caused by herniated disc compression of nerve roots, we wondered to what extent this form of proximal nerve crush injury was associated with spinal cord neuronal apoptosis and whether exogenous rhEPO therapy might reduce apoptosis in conjunction with attenuations in neuropathic pain.

There are several nerve root injury models that are used to induce mechanical allodynia. We chose nerve root crush; this type of injury induces a mechanical allodynia only in the ipsilateral hind paw (Kawekami et al., 1994). Other nerve root injury models that use chromic gut ligatures manifest mechanical allodynia in both ipsilateral and contra lateral hind paws (Hunt et al., 2001). Nerve root crush resulted in eighty percent of the animals manifesting pain behaviors. However, in rhEPO treated animals, pain behaviors resolved much more quickly (one week sooner) than vehicle treated animals. Similar to the results of the L5 SNC, rhEPO treatment was not initially protective against acute pain states, suggesting that EPO stimulates intracellular signaling cascades and gene transcription, stabilizing later changes after nerve injury. As suggested previously, rhEPO may have a key role in nerve regeneration. In these studies, rhEPO was given prior to the injury which may limit the potential clinical value of rhEPO treatment as a therapy for lumbar radiculopathy associated with disc prolapse. However, many patients with sciatica or low back pain may receive rhEPO as a preemptive treatment to surgical procedures. In addition, studies have now shown the efficacy of rhEPO treatment after injury in both spinal cord and retina (Celik et al., 2002; Junk et al., 2002).

Injury to nerve roots causes profound changes in input from primary afferents innervating the dorsal horn. Some of those changes include increased neurotransmitter release; for example, the excitatory neurotransmitter, glutamate, is released leading to excitotoxicity and death of spinal dorsal horn neurons. We showed that vehicle-treated rats receiving a nerve root crush injury induced apoptosis of spinal dorsal horn neurons on ipsilateral (day 1) and both ipsilateral and contralateral (day 3) sides. In contrast, rhEPO treated rats had significantly less apoptotic spinal dorsal horn neurons in both ipsilateral and contralateral sides (Sekiguchi et al., 2003). One plausible mechanism is that rhEPO protected neurons from glutamate excitotoxicity in spinal dorsal horn neurons. It has been reported that EPOR-mediated inhibition of excitotoxic glutamate release conferred neuroprotection to cerebellar granule neurons (Kawakami et al., 2001). Direct effects of EPO on spinal dorsal horn neurons are plausible since receptors for EPO are subcellularly distributed on their membranes (Sekiguchi et al., 2003) and rhEPO crosses the blood-brain barrier (Brines et al., 2000). Alternatively, and as suggested by the CCI data, rhEPO may be neuroprotective indirectly, by downregulating TNF-α, a key pain producing cytokine that is secreted from ruptured discs (Igarashi et al., 2000) and is downregulated by rhEPO (Villa et al., 2003; Campana et al., 2004).

4. EFFECTS OF EPO DURING METABOLIC INJURY: DIABETIC NEUROPATHY

More than half of all patients with diabetes mellitus develop some form of peripheral neuropathy. The most common presentation is a degenerative distal symmetrical polyneuropathy with progressive sensory loss, while around 10% of patients also describe concomitant neuropathic pain. There are currently no approved treatments that prevent or reverse degenerative diabetic neuropathy and although the neuropathic pain of some patients may be alleviated by agents such as gabapentin or tramadol, these drugs do not address the underlying etiology of the disorder.

Diabetic rodents model a number of the functional and early structural disorders associated with diabetic neuropathy and are widely used to investigate potential etiologic mechanisms and to screen novel therapeutic approaches for preventing degenerative neuropathy and alleviating neuropathic pain. In a recent study, systemic delivery of rhEPO was shown to prevent and reverse a variety of nerve disorders present in insulin-deficient diabetic rats (Bianchi et al., 2004). These included conduction slowing, a functional index of early degenerative neuropathy in large myelinated fibers, and thermal hypoalgesia, an index of small fiber mediated sensory loss. Loss of epidermal fibers, a structural correlate to thermal hypoalgesia, was also ameliorated by rhEPO. In addition to these actions on indices of degenerative neuropathy, rhEPO also ameliorated mechanical hyperalgesia in diabetic rats, suggesting a therapeutic potential for rhEPO in treating both the degenerative and painful manifestations of diabetic neuropathy.

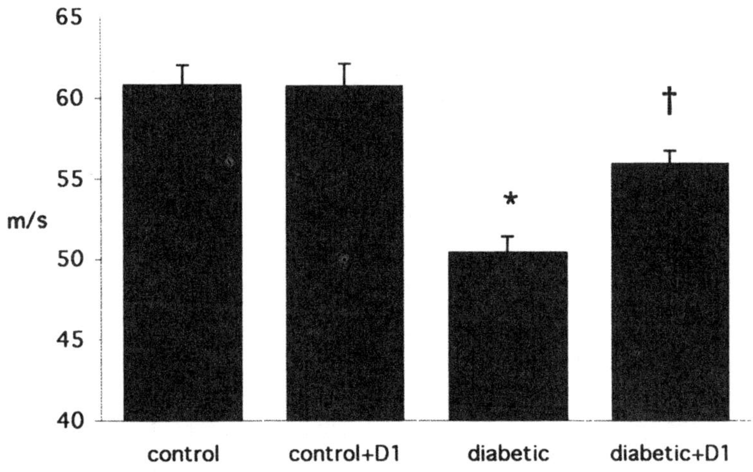

Figure 10-1. Sensory nerve conduction velocity in the sciatic nerve of control and streptozotocin-induced diabetic rats after 4 weeks of treatment with peptide D1 (200 μg/kg/day) or vehicle. Data are mean ± SEM of N=9-10 rats per group. Statistical comparisons by one-way ANOVA with Student-Newman-Keuls post hoc test. * = $P<0.01$ vs. all other groups; † = $p<0.05$ vs. all other groups.

One caveat to the promising findings with rhEPO treatment in diabetic rats is the observation that hematocrit increased by 45% within 5 weeks of starting treatment (Bianchi et al., 2004). This would likely limit the therapeutic value of the holoprotein in diabetic patients and other approaches that emulate neuroprotective properties of rhEPO, without inducing hematopoeisis, may be more rewarding. Using an approach previously applied to identifying the neuroactive region of the prosaposin molecule (O'Brien et al., 1995) that was neuroprotective in diabetic rats (Calcutt et al., 1999), we identified the regions of the EPO molecule that possess neuroactive properties (Campana et al., 1998). Treating insulin-deficient diabetic rats with a protected sequence of a neuroactive region of EPO (peptide D1) ameliorated conduction slowing of large sensory fibers

(Campana and Calcutt, unpublished observations: Fig. 10-1). Moreover, daily treatment with peptide D1 for 8 weeks did not increase hematocrit in either control or diabetic rats, and there were no effects on the severity of diabetes as reflected by loss of body weight and hyperglycemia. These findings suggest that selective targeting of the EPO receptor in peripheral nerve is possible and that it may provide a novel approach to treating diabetic, and plausibly other, peripheral neuropathies.

5. CONCLUSIONS AND FUTURE DIRECTIONS

Damage to peripheral nerves underlies many neurodegenerative and chronic neuropathies including diabetic neuropathy, AIDS neuropathy, post-herpetic neuralgia, chemotherapy and radiation-induced neuropathies, and traumatic nerve injuries. Neuropathic pain is both prevalent and resistant to current pharmacological therapy, which includes opioids, anti-convulsants, NSAIDS and local anesthetics. Understanding the cellular and molecular mechanisms underlying these peripheral disorders has helped identify appropriate targets for drug discovery. One novel mechanism is the prevention of apoptosis of peripheral and central neurons and their supporting glia following nerve damage. Developing rhEPO, a known neuroprotectant, will be of value, particularly when peptides or mimetics are developed that facilitate the neuroprotective aspect of rhEPO without the hematological side effects (Campana et al., 1998; Leist et al., 2004).

REFERENCES

Bennett, G. J., Xie, Y.-K., 1988, A peripheral mononeuropathy in rat that produces disorders of pain sensation like those seen in man, *Pain* **33:**87-107.

Bernaudin, M., Bellail, A., Marti, H. H., Yvon, A., Vivien, D., Duchatelle, I., Mackenzie, E. T., Petit, E., 2000, Neurons and astrocytes express Epo mRNA: Oxygen-sensing mechanisms that involve the redox state of the brain, *GLIA* **30:**271-278.

Bianchi, R., Buyukakilli, B., Brines, M., Savino, C., Cavaletti, G., Oggioni, N., Lauria, G., Borgna, M., Lombardi, R., Cimen, B., Comelekoglu, U., Kanik, A., Tataroglu, C., Cerami, A., Ghezzi, P., 2004, Erythropoietin both protects from and reverses experimental diabetic neuropathy, *PNAS* **101:**823-828.

Bittorf, T., Jaster, R., Ludke, B., Kamper, B., Brock, J., 1997, Requirement for JAK2 in erythropoietin-induced signaling pathways, *Cell Signal.* **9:**85-89.

Braczkowski, R., Romanawsky, W., Danikiewicz, A., Muc-Wierzgon, M., Blazelonis, A., Zubelewicz, B., 2001 Decrease of erythropoietin level by human recombinant tumour necrosis factor alpha (hrec TNFalpha) in patients with advance cancer. *J Biol Regul Homeost Agents* 15:366-9.

Brines, M. L., Ghezzi, P., Keenan, S., Agnello, D., de Lanerolle, N. C., Cerami, C., Itri, L. M., Cerami, A., 2000, Erythropoietin crosses the blood-brain barrier to protect against experimental brain injury, *PNAS* **97:**10526-10531.

Calcutt, N.A. Campana, W.M., Eskeland N.L., Mohiuddin L., Dines K.C., Mizisin A.P., O'Brien, J.S., 1999, Prosaposin gene expression and the efficacy of a prosaposin-derived inpreventing structural and functional disorders of peripheral nerve in diabetic rats, *J. Neuropath. Expt. Neurol.* 58:628-636.

Campana, W. M., Misasi, R., O'Brien, J. S., 1998, Identification of a neurotrophic sequence in erythropoietin, *Int. J. Mol. Med.* **1:**235-241.

Campana, W. M., Myers, R. R., 2001, Erythropoietin and erythropoietin receptors in the peripheral nervous system: Changes after nerve injury, *FASEB J.* 10.1096/fj.00-0857fje published online June 8, 2001.

Campana, W. M., Myers, R. R., 2003, Exogenous erythropoietin protects against dorsal root ganglion apoptosis and pain following peripheral nerve injury, *Eur. J. Neurosci.* **18:**1497-1506.

Campana, W.M., Angert, M., Shubayev, V., Myers, R.R., 2004, Erythropoietin reduces Schwann cells TNF-α expression and pain after peripheral nerve injury. 23rd Annual Meeting of the American Pain Society, Vancouver, Canada.

Celik, M., Gokmen, N., Erbayraktar, S., Akhisaroglu, M., Konak, S., Ulukus, C., Genc, S., Genc, K., Sagiroglu, E., Cerami, A., Brines, M., 2002, Erythropoietin prevents motor neuron apoptosis and neurologic disability in experimental spinal cord ischemic injury, *PNAS* **99:**2258-2263.

Chaplan, S. R., Pogrel, J. W., Yaksh, T. L., 1994, Role of voltage dependent calcium channel subtypes in experimental tactile allodynia, *J. Pharmacol. Exp. Ther.* **269:**1117-1123.

Cooper, A.C., Mikhail, A., Lethbridge, M.W., Kemeny, D.M., Macdougall, I.C., 2003, Increased expression of erythropoiesis inhibiting cytokines (IFN-γ, TNF-α, IL-10 and IL-3) by T cells inpatients exhibiting a poor response to erythropoietin therapy, *J. Am. Soc. Nephrol.* 14: 1776-1784.

Deckworth, T.L., Johnson, E.M., 1993, Neurotrophic factor deprivation-induced death, Ann. N. Y. Acad. Sci. **28:**121-131.

Degn, J., Tandrup, T., Jakobsen, J., 1999, Effect of nerve crush on perikaryal number and volume of neurons in adult dorsal root ganglion, *J. Comp. Neurol.* **412**:186-192.

Digicaylioglu, M., Bichet, S., Marti, H. H., Wenger, R. H., Rivas, L. A., Bauer, C., Gassmann, M., 1995, Localization of specific erythropoietin binding sites in defined areas of the mouse brain, *PNAS* **92:**3717-3720.

Digicaylioglu, M., Lipton, S. A., 2001, Erythropoietin-mediated neuroprotection involves cross-talk between Jak2 and NF-kB signaling cascades, *Nature* **412:**641-647.

Hargreaves, K., Dubner, R., Brown, F., Flores, C., Joris, J., 1988, A new and sensitive method for measuring thermal nociception in cutaneous hyperalgesia, *Pain* **32:**77-88.

Hunt, J.L., Winkelstein, B.A., Rutkowski, M.D., Weinstein, J.N., De Leo, J.A., 2001, Repeated injury to the lumbar nerve roots produces enhanced mechanical allodynia and persistent spinal neuroinflammation, *Spine,* **26**:2073-9.

Igarashi, T., Kikuchi, S., Shubayev, V., Myers, R.R., 2000, Volvo Award winner in basic science studies: Exogenous tumor necrosis factor-alpha mimics nucleus pulposus-induced neuropathology. Molecular, histologic, and behavioral comparisons in rats, *Spine,* **25**:2975-80.

Junk, A.K., Mammis, A., Savitz, S.I., Singh, M., Roth, S., Malhotra, S., Rosenbaum, P.S., Cerami, A., Brines, M., Rosenbaum, D.M., 2002, Erythropoietin administration protects retinal neurons from acute ischemia-reperfusion injury, *Proc. Natl. Acad. Sci. U.S.A.*, **99**: 10659-64.

Kawakami, M., Sekiguchi, M., Sato, K., Kozaki S., Takahashi, M., 2001, Erythropoietin receptor-mediated inhibition of excytotoxic glutamate release confers neuroprotection during chemical ischemia, *J. Biol. Chem.* **276**:39469-75.

Kawakami, M., Weinstein, J. N., Chatani, K., Spratt, K. F., Meller, S.T., Gebhart, G.F., 1994, Experimental lumbar radiculopathy. Immunohistochemical and quantitative demonstrations of pain induced by lumbar nerve root irritation of the rat, *Spine* **19:**1780-1794.

Konishi, Y., Chui, D.-H., Hirose, H., Kunishita, T., Tabira, T., 1993, Trophic effect of erythropoietin and other hematopoietic factors on central cholinergic neurons in vitro and in vivo, *Brain Res.* **609:**29-35.

Leist, M., Ghexxi, P., Grasso, G., Bianchi, R., Villa, P., Fratelli, M., Savino, C., Bianchi, M., Nielson, J., Gerwien, J., Kallunki, P., Larsen, A.-K., Helboe, L., Christensen, S., Pedersen, L., Nielson, M., Torup, L., Sager, T., Sfacteria, A., Erbayraktar, S., Erbayraktar, Z., Gokmen, N., Yilmaz, O., Cerami-Hand, C., Xie, Q., Colman, T., Cerami, A., Brines, M., 2004, Derivatives of erythropoietin that are tissue protective but not erythropoietic, *Science* **305:**239-242.

Li, X., Gonias, W.M., and Campana W.M., 2005, Schwann cells express erythropoietin receptor and represent a major target for erythropoietin in peripheral nerve injury, *Glia* **51:**254-65.

Mao, J., Sung, B., Ji, R. R., Lim, G., 2002, Neuronal apoptosis associated with morphine tolerance: Evidence for an opioid-induced neurotoxic mechanism, *J. Neurosci.* **22:**7650-7661.

Meakin, S. O., Shooter, E. M., 1992, The nerve growth factor family of receptors, *Trends Neurosci.* **15:**323-331.

Miura, Y., Miura, O., Ihle, J. N., Aoki, N., 1994, Activation of the mitogen activated protein kinase pathway by the erythropoietin receptor, *J. Biol. Chem.* **269:**29962-29969.

Moore, K.A., Kohno, T., Karchewski, L.A., Scholz, J., Baba, H., Woolf, C.J., 2002, Partial peripheral nerve injury promotes a selective loss of GABAergic inhibition in the superficial dorsal horn of the spinal cord. *J. Neurosci.* **22**: 6724-31.

Myers, R. R., Yamamoto, T., Yaksh, T. L., Powell, H. C., 1993, The role of focal nerve ischemia and Wallerian degeneration in peripheral nerve injury producing hyperesthesia, *Anesthesiology* **78:**308-316.

Myers, R. R., 1998, Morphology of the peripheral nervous system and its relationship to neuropathic pain, in: *Anesthesia: Biologic Foundations,* T. L. Yaksh, C. Lynch III, W. M. Zapol, M. Maze, J. F. Biebuyck, L. J. Saidman, ed., Lippincott-Raven, Philadelphia, pp. 483-514.

O"Brien, J.S., Carson, G.S., Seo, H.C., et al., 1995, Identification of the neurotrophic factor sequence of prosaposin, *FASB J,* **9:**681-85

Qiu, J., Nesic, O., Ye, Z., Rea, H., Westlund, K. N., Xu, G. Y., McAdoo, D., Hulsebosch, C. E., Perez-Polo, J. R., 2001, Bcl-XL expression after contusion to the rat spinal cord, *J. Neurotrauma* **18:**1267-1278.

Sakanaka, A., Wen, T.-C., Matsuda, S., Masuda, S., Morishita, E., Nagao, M., Sasaki, R., 1998, In vivo evidence that erythropoietin protects neurons from ischemic damage, *PNAS* **95:**4635-4640.

Sekiguchi, Y., Kikuchi, S., Myers, R. R., Campana, W. M., 2003, ISSLS Prize Winner: Erythropoietin inhibits spinal neuronal apoptosis and pain following nerve root crush, *Spine* **28:**2577-2584.

Shu, X. Q., Mendell, L. M., 1999, Neurotrophins and hyperalgesia, *Proc. Natl. Acad. Sci.* **96:**7693-7696.

Sirén, A.-L., Fratelli, M., Brines, M., Goemans, C., Casagrande, S., Lewczuk, P., Keenan, S., Gleiter, C., Pasquali, C., Capobianco, A., Mennini, T., Heumann, R., Cerami, A., Ehrenreich, H., Ghezzi, P., 2001, Erythropoietin prevents neuronal apoptosis after cerebral ischemia and metabolic stress, *PNAS* **98:**4044-4049.

Sorkin, L.S., Xiao, W.H., Wagner, R., Myers, R.R., 1997, Tumour necrosis factor-alpha induces ectopic activity in nociceptive primary afferent fibres, *Neuroscience,* **81**:255-62.

Vairano, M., Dello Russo, C., Pozzoli, G., Battaglia, A., Scambia, G., Tringali, G., Aloe-Spiriti, M. A., Preziosi, P., Navara, P., 2002, Erythropoietin exerts anti-apoptotic effects on rat microglial cells in vitro, *Eur. J. Neurosci.* **16:**684-692.

Villa, P., Bigini, P., Mennini, T., Agnello,D., Laragione, T., Cognotto, A., Viviani, B., Marinovich, M., Cerami, A., Coleman, T.R., Brines, M., Ghezzi, P., 2003, Erythropoieitn selectively attenuates cytokine production and inflammation in cerebral ischemica by targeting neuronal apoptosis, *J. Exp. Med.* 198:971-975.

Wagner, R., Myers, R. R., 1996, Schwann cells produce tumor necrosis factor alpha: Expression in injured and non-injured nerves, *Neuroscience* **73:**625-629.

Weishaupt, J. H., Rohde, G., Polking, E., Siren, A.-L., Ehrenreich, H., Bahr, M., 2004, Effect of erythropoieitn axotomy-induced apoptosis in rat retinal ganglion cells, *Invest. Ophthalmology & Visual Sci.* **45:**1514-1522.

Whiteside, G. T., Munglani, R., 2001, Cell death in the superficial dorsal horn in a model of neuropathic pain, *J. Neurosci. Res.* **64:**168-173.

Winkelsein, B.A., Rutowski, M.D., Sweitzer, S.M., Pahl, J.L., De Leo, J.A., 2001, Nerve injury proximal or distal to the DRG induces similar spinal glial activation and selective cytokine expression but differential behavioral responses to pharmacologic treatment., J. Comp Neurol. **439**:127-39.

Chapter 11

AN ENDOGENOUS PATHWAY PREVENTING AXONAL DEGENERATION MEDIATED BY SCHWANN CELL – DERIVED ERYTHROPOIETIN

Sanjay C. Keswani and Ahmet Höke

Department of Neurology, Johns Hopkins University School of Medicine, Baltimore, MD, USA

Abstract: Most common peripheral neuropathies are characterized by distal axonal degeneration, rather than neuronal death. The pathways that mediate axonal degeneration and neuronal death are likely to be diverse. In this chapter, we describe a novel, endogenous pathway that prevents axonal degeneration. We show that in response to axonal injury, peri-axonal Schwann cells release erythropoietin (EPO), which via binding to EPO receptors on neurons prevents axonal degeneration. We demonstrate that the relevant axonal injury signal that stimulates EPO production from surrounding glial cells is nitric oxide. In addition, we show that this endogenous pathway can be therapeutically exploited by administering exogenous EPO in vivo and in vitro. Our data suggest that EPO prevents axonal degeneration, and may therefore be therapeutically useful in a wide variety of human neurological diseases characterized by axonopathy.

Key words: axonal degeneration, Schwann cells, neuroprotection, neuropathy, axonopathy

1. INTRODUCTION

Peripheral neuropathies are common and cause significant morbidity. The vast majority of peripheral neuropathies, including diabetic and HIV-associated neuropathy, are 'dying back' axonopathies, characterized by degeneration of the most distal portions of axons, with centripetal progression (Sidenius, 1982; Pardo et al., 2001). Although most published in vitro studies of neurotoxicity and neuroprotection in the peripheral nervous system have focused on neuronal apoptosis as the sole outcome measure, neuronal death, in contrast to distal axonal loss, is not a prominent pathological feature of most human peripheral neuropathies. Furthermore, the signaling pathways mediating axonal degeneration are distinguishable from those mediating neuronal apoptosis (Glass et al., 2002; Raff et al., 2002; Zhai et al., 2003; Ehlers, 2004). Thus, in considering whether a particular 'neuroprotective' agent may have therapeutic relevance to human peripheral neuropathies (and to other neurological diseases where axonopathy is prominent), it is important to discover if it robustly prevents axonal degeneration, independent of neuronal death (Coleman and Perry, 2002).

2. EPO NEUROPROTECTION IN THE PNS

The glycoprotein, erythropoietin (EPO), is a very promising neuroprotective agent, whose anti-apoptotic properties have been thoroughly evaluated by several investigators. The administration of EPO prevents central nervous system neurons from death caused by a variety of insults, including hypoxia, hypoglycemia, glutamate toxicity, growth factor deprivation and free radical injury (Digicaylioglu and Lipton, 2001; Siren et al., 2001; Chong et al., 2002; Gorio et al., 2002; Ruscher et al., 2002). Recently, Campana et al also demonstrated that EPO administration prevented apoptosis of DRG sensory neurons (Campana and Myers, 2003). In this chapter, we discuss the ability of EPO to prevent axonal degeneration in the PNS. Furthermore, we demonstrate the evidence for an endogenous 'axonoprotective' pathway mediated by EPO production from Schwann cells, the major glial cells of the PNS (Keswani et al., 2004).

2.1 EPO and EPOR expression in DRG neurons and Schwann cells

Immunostaining of dissociated DRG neuron-Schwann cell co-cultures reveal that both neurons and Schwann cells express EPO (Fig. 11-1A), whereas neurons predominantly express EPO-R (Fig. 11-1B). Of interest, as can be seen in Figure 1b, neuronal EPO-R is localized on axons as well as perikarya. A similar pattern of EPO and EPO-R immunostaining is observed in DRG sections harvested from adult rats (Figs. 11-1C and D), EPOR immunostaining again being particularly intense in DRG neurons as compared to Schwann cells.

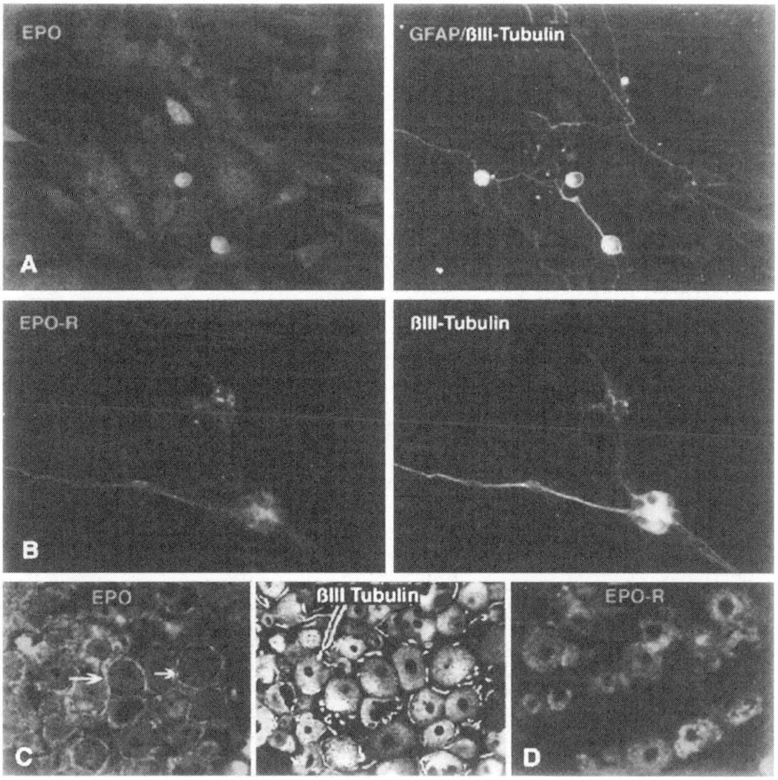

Figure 11-1. In vitro and vivo EPO and EPOR expression by DRG neurons and Schwann cells: (A) Triple immunoflouresent labeling of dissociated DRG neuron- Schwann cell co-cultures shows that EPO immunostaining is present in both neurons (BIII-tubulin-labeled) and Schwann cells (GFAP-labeled). (B) EPOR is present in axons and cell bodies of DRG neurons in the co-cultures (C) Immunostaining of adult rat DRG sections shows that EPO is expressed by DRG neurons and peri-neuronal Schwann cells (arrows). (D) In contrast, similar to the in vitro staining, EPOR immunoreactivity (red) is mainly in DRG neurons, rather than Schwann cells. Scale bars = 50μm (Adapted from Keswani et.al., Annals of Neurology 2004).

2.2 Neighboring Schwann cells produce EPO in response to axonal injury

Following transaction of axons in well-established, dissociated DRG cultures, EPO immunostaining is markedly increased in Schwann cells that are in close proximity to the transected axons as compared to those adjacent to uninjured axons (Fig. 11-2) (Keswani et al., 2004). Furthermore, following unilateral sciatic nerve transection in adult rats, EPO mRNA levels are increased four to five-fold in both the sciatic nerve as well as the lumbar DRG harvested from the cut side as compared to the contralateral non-cut side (Fig. 2B). As sciatic nerve does not contain neuronal mRNA – there are no neuronal cell bodies in peripheral nerve – it is likely that the increased EPO mRNA production occurred in Schwann cells rather than in neurons. This correlates with a study by Campana et al (Campana and Myers, 2001), which showed that Schwann cells in peripheral nerve have increased EPO immunostaining following in vivo axotomy. Of interest, mRNA levels for EPOR, are increased in lumbar DRG (but not in sciatic nerve) harvested from the transected side, being 3.5 and 2.5 fold higher than the control side at 4 and 24 hours respectively.

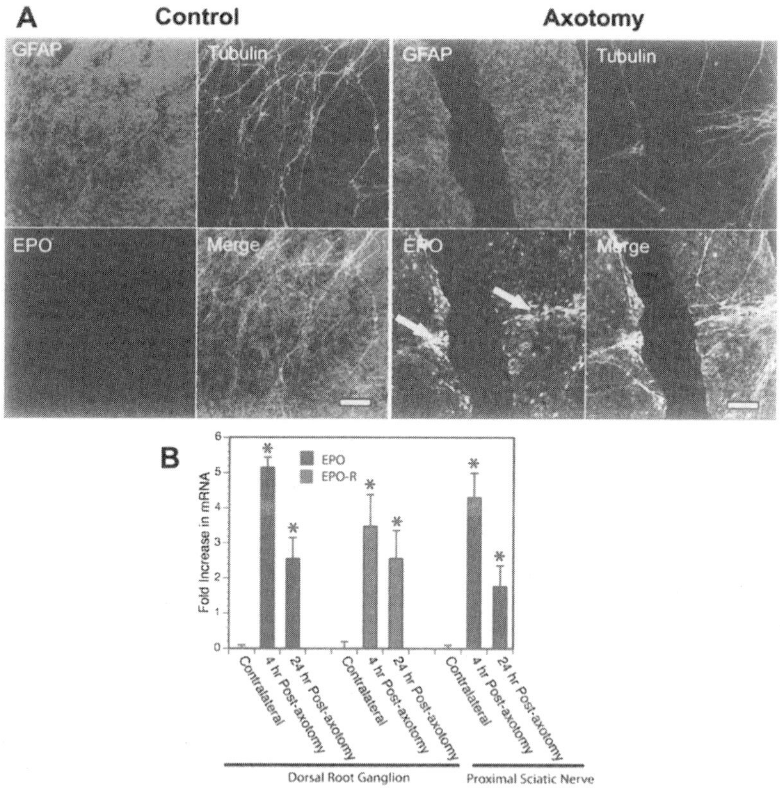

Figure 11-2. Neighboring Schwann cells produce EPO in response to axonal injury. (A) Axotomy in DRG neuron – Schwann cell co-cultures results in a marked increase at 4 hours in the expression of EPO in Schwann (GFAP-labeled) cells neighboring transected axons (ßIII-tubulin-labeled). Scale bar 100μm. (B) Semi-quantitative real-time PCR for rat EPO shows a robust increase in EPO mRNA levels in both the Lumbar DRG and the proximal sciatic nerve at 4 and 24 hours post axotomy at the mid-sciatic level. Accompanying this, there is a more modest increase in EPOR mRNA in the lumbar DRG after sciatic nerve transection. (Adapted from Keswani et.al., Annals of Neurology 2004)

To explore other axonal injury paradigms, dissociated DRG cultures were exposed to neurotoxins at doses known to reproducibly cause 'dying back' axonal degeneration but not neuronal death. The neurotoxins chosen were gp120, the HIV envelope glycoprotein, and ddC, an antiretroviral agent known to cause peripheral neuropathy. The neurotoxicity profiles of these agents have been well characterized in our culture system (Keswani et al., 2003a; Keswani et al., 2003b). After 24 hours of exposure to gp120 (1 pg/ml) or ddC (10 μM), EPO immunostaining in peri-neuronal Schwann cells was greatly increased compared to vehicle control - treated cultures (Keswani et al., 2004). To investigate if there was extracellular EPO release, the supernatants of gp120, ddC and vehicle control-treated DRG cultures were analyzed by Western blotting. EPO protein was increased in the supernatants of neurotoxin-treated cultures compared to that of vehicle control - treated cultures. In contrast, when gp120 or ddC was applied to pure Schwann cell cultures, no EPO induction was noted by EPO immunostaining or Western blotting of supernatants (Keswani et al., 2004). These findings coupled with our observations that after neurotoxin exposure or axonal transection, only those Schwann cells in intimate contact with axons had increased EPO immunostaining, suggest that 'sick axons' are needed in close proximity to Schwann cells for EPO induction in those Schwann cells to occur.

2.3 Axonal injury stimulates Schwann cell production of EPO via nitric oxide

What is the identity of the neuronal/axonal 'injury factor' that stimulates EPO production by neighboring Schwann cells? We screened a number of promising candidates, including neuregulin-1 and Insulin Growth factor-1 (IGF-1), without success. Finally, we discovered that nitric oxide might be the relevant signaling molecule, on the basis of the following observations. All the agents that we noted had caused dying back axonal degeneration in our cultures, including gp120, ddC and acrylamide, increased neuronal intracellular NO production (Fig 11-3A). This observation correlates with previous studies showing that neuronal nitric oxide synthase (nNOS) gene expression is significantly increased at 4 hours in ipsilateral DRG samples following sciatic nerve injury in a rat tourniquet model (Mizusawa et al., 2003). We also noted that NO donors, such as SNAP and NOR-3, increased EPO mRNA levels in pure Schwann cell cultures as early as 30 minutes after administration, with a three- to four-fold increase being noted at 1 hour (Fig. 11-3B). This was mirrored by an increase in intracellular EPO production by Western blotting (Fig. 11-3C) and a large increase in EPO content in the supernatants of these cultures, as measured by ELISA (Fig. 11-3D). Co-administration of L-NAME, a non-specific NOS inhibitor, almost completely obliterated the ability of gp120 to induce EPO release into the supernatants of DRG cultures (Fig. 11-3E). Moreover, TRIM, a specific nNOS inhibitor (Haga et al., 2003), completely prevented the 18-fold induction of EPO mRNA by gp120 in these cultures, suggesting that NO generated by nNOS was responsible for triggering EPO production by surrounding glial cells. In our dissociated DRG cultures, immunostaining for nNOS only occurred in neurons, in contrast to iNOS staining which was present in both neurons and Schwann cells.

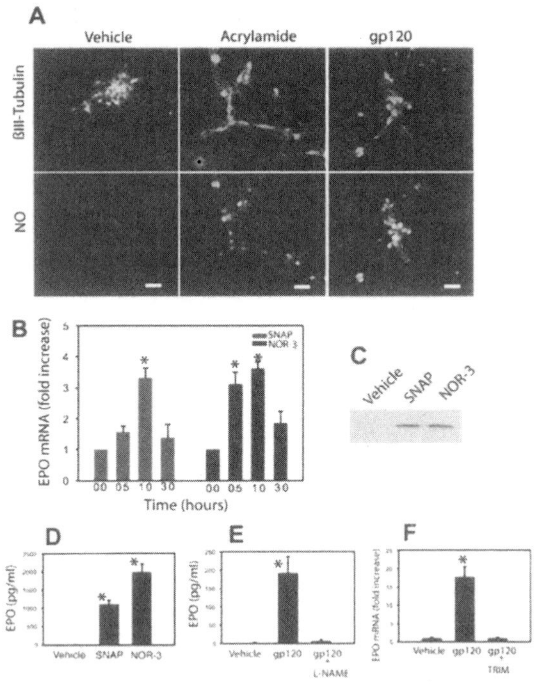

Figure 11-3. Axonal injury stimulates Schwann cell production of EPO via Nitric Oxide. (A) Exposure of DRG co-cultures to agents causing axonal degeneration, including gp120 (1pg/ml) and acrylamide (1mM), induces NO production at 6 hours in ßIII tubulin labeled neurons, as assayed by DAF-2T fluorescence. Scale bar 100μm. (B) NO donors, SNAP (10 μM) and NOR-3 (100 nM), induce increased EPO mRNA levels ($p<0.05$) in pure Schwann cell cultures. (C) Cell lysates of pure Schwann cell cultures treated for 6 hours with SNAP (10μM) or NOR-3 (100nM), have increased EPO protein by Western blotting, compared to those treated with vehicle control. (D) Supernatants of pure Schwann cell cultures treated for 24 hours with SNAP (10μM) or NOR-3 (100nM) have markedly increased EPO content by ELISA, compared to vehicle control treatment (* $p<0.05$). (E) L-NAME (100μM) co-administration prevents gp120-induced EPO release from DRG neuron - Schwann cell co-cultures, as measured by EPO ELISA (* $p<0.05$). (F) TRIM (100μM) co-administration abrogates gp120-induced (18-fold) increase in Schwann cell EPO mRNA in DRG neuron-

Schwann cell co-cultures (* $p<0.05$). (Adapted from Keswani et.al., Annals of Neurology 2004)

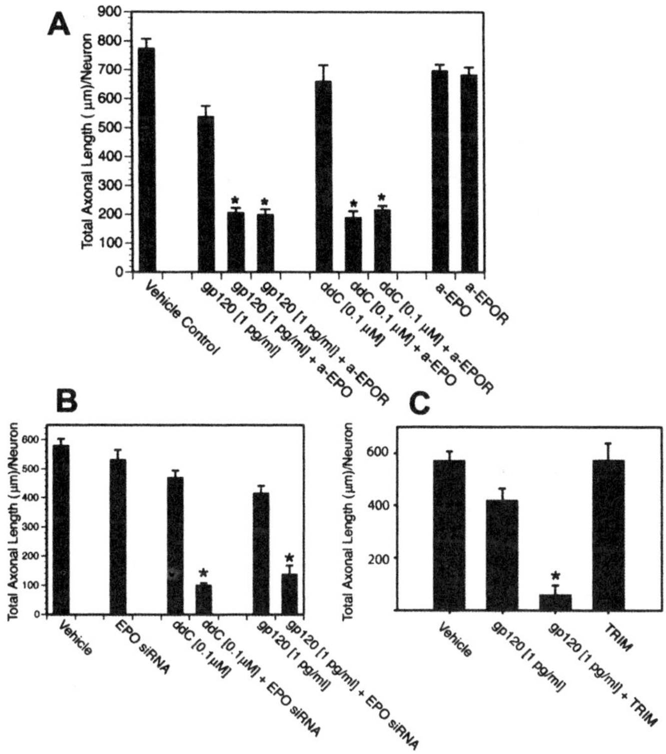

Figure 11-4. Endogenous EPO production by Schwann cells is 'axonoprotective'. (A) The application of antibodies against EPO or EPO-R increases the sensitivity of DRG axons to ddC (0.1 µM) and gp120 (1pg/ml) –induced degeneration. (* $p<0.05$). (B) Downregulation of endogenous erythropoietin expression in Schwann cells within DRG co-cultures by EPO siRNA increases axonal degeneration induced by ddC and gp120 (* $p<0.05$). (C) Similarly, co-administration of TRIM (100µM) to DRG co-cultures increases gp120-induced axonal degeneration (* $p<0.05$). (Adapted from Keswani et.al., Annals of Neurology 2004)

2.4 Schwann cell-derived EPO prevents axonal degeneration

What is the relevance of this Schwann cell derived EPO? When EPO gene silencing in Schwann cells was performed by transfection with anti-EPO siRNA, DRG axons were noted to be far more vulnerable to degeneration by ddC and gp120 (Fig 11-4) (Keswani et al., 2004). This was associated with a lack of EPO induction in peri-axonal Schwann cells by ddC and gp120 in the transfected cultures. The 'axonoprotective' efficacy of endogenous EPO was further suggested by similarly increased axonal degeneration by ddC and gp120 when antagonist antibodies to EPO or to EPOR were co-administered. No associated increase in neuronal death was observed by ethidium homodimer staining (which would detect both apoptotic and necrotic death), suggesting that the increased axonal degeneration by EPO/EPOR antagonism was not due to neuronal death.

2.5 Endogenous Nitric Oxide prevents axonal degeneration

Similar to the effect of endogenous EPO antagonism in the DRG co-cultures, the application of TRIM, a specific nNOS inhibitor, resulted in markedly increased axonal degeneration induced by gp120 (1pg/ml) (Fig. 11-Cd). As in previous experiments, TRIM co-administration with gp120 did not result in increased neuronal death. These findings, in combination with our previous observations (Fig.11-3), suggest the importance of nNOS in the endogenous EPO 'axonoprotective' response to axonal injury. The dual role of NO with respect to neurotoxicity and neuroprotection has been commented on in the literature (Wiggins et al., 2003; Bolanos et al., 2004). While NO-mediated neurotoxicity has been explored by several groups over the years, NO-mediated neuroprotection is poorly understood. In our study, nNOS inhibition exacerbates the axonal degeneration induced by gp120. This correlates with prevention by TRIM of Schwann cell- derived EPO production in response to axonal injury. Of some relevance to this discussion is a study by Keihoff et al, which showed that nNOS knockout mice had worsened axonal degeneration following sciatic nerve transaction compared to wild type mice (Keilhoff et al., 2002). Figure 11-5 summarizes our model of endogenous 'axonoprotection' by Schwann cell –derived EPO.

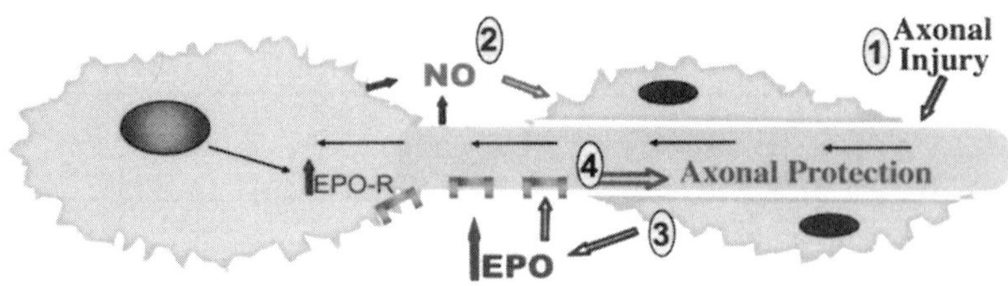

Figure 11-5. Schematic diagram of EPO-mediated intrinsic axonoprotective pathway. Based on our data, we hypothesize the following: Axonal injury (1) induces nitric oxide production (2) within neurons. This neuron-derived NO stimulates EPO production (3) by neighboring Schwann cells. This Schwann cell-derived EPO results in activation of an 'axonoprotective' pathway (4) via EPOR ligation on neurons.

2.6 Recombinant human EPO prevents axonal degeneration in models of peripheral neuropathy

Can this endogenous 'axonoprotective' pathway be therapeutically exploited by systemic administration of EPO? This was tested in a well-characterized animal model of peripheral axonal degeneration, namely the rat acrylamide toxicity model (Gold et al., 1985; LoPachin and Lehning, 1994; Crofton et al., 1996; Ko et al., 1999). In this model, oral acrylamide administration to Sprague-Dawley rats results in severe 'dying back' degeneration of both sensory and motor fibers, in the absence of significant neuronal death (Fullerton and Barnes, 1966; LoPachin and Lehning, 1994). Affected rats characteristically have distal limb weakness and an ataxic gait (DeGrandchamp and Lowndes, 1990; Ko et al., 1999). Furthermore, we noted that acrylamide-treated rats developed mechanical hyperalgesia, a correlate of neuropathic pain behavior. In this study (Keswani et al., 2004), acrylamide-treated rats given EPO had significantly less sensory axonal degeneration as indicated by greater cutaneous innervation (increased epidermal nerve fiber density) on pgp 9.5 immunohistochemistry, compared to those given placebo. This correlated with decreased mechanical hyperalgesia on Von Frey filament testing. Furthermore, EPO-treated rats had significantly less motor axonal degeneration as demonstrated by a higher innervated neuromuscular junction density in the intrinsic foot muscles using α-bungaratoxin binding. This correlated with greater grip strength.

3. CONCLUSION

Progressive 'dying back' degeneration of the distal regions of long axons, rather than neuronal loss, is the predominant pathological change in the most common peripheral neuropathies afflicting humans, such as diabetic sensorimotor polyneuropathy, HIV-associated sensory neuropathy and toxic neuropathies (Sidenius, 1982; Gold et al., 1985; Pardo et al., 2001; Keswani et al., 2002). Furthermore, progressive axonal loss is observed in multiple sclerosis, and is now thought to highly correlate with disability (Bjartmar and Trapp, 2001; Bjartmar et al., 2003). Consequently, agents with 'axonoprotective' properties may be very helpful therapeutically. However, often only the anti-apoptotic properties of putative neuroprotective agents are evaluated, with little or no attention paid to whether axonal degeneration can be prevented. It does not necessarily follow that an agent that prevents neuronal apoptosis will prevent axonal degeneration, as it is now well recognized that the two processes may exploit different signaling pathways

(Glass et al., 2002; Raff et al., 2002; Zhai et al., 2003; Ehlers, 2004; Korhonen and Lindholm, 2004).

In this chapter, we describe an endogenous EPO-mediated pathway that prevents axonal degeneration. We show that in response to axonal injury, peri-axonal Schwann cells release EPO, which via EPO-R binding on neurons prevents axonal degeneration. We demonstrate that the relevant axonal injury signal that stimulates EPO production from surrounding glial cells is nitric oxide. In addition, we show that this endogenous pathway can be therapeutically exploited by administering exogenous EPO. In an animal model of peripheral neuropathy, systemic EPO administration prevents axonal degeneration, and this is associated with a reduction in limb weakness and neuropathic pain behavior. Our in vivo and in vitro data suggest that EPO prevents axonal degeneration, and may therefore be therapeutically useful in a wide variety of human neurological diseases characterized by axonopathy.

ACKNOWLEDGEMENTS

This work was supported by grants from the National Institutes of Health (NS43991, NS46262, NS47972), the Center for AIDS Research, Johns Hopkins University, and by a contract from the R. W. Johnson Pharmaceutical Research Institute.

REFERENCES

Bjartmar C, Trapp BD (2001) Axonal and neuronal degeneration in multiple sclerosis: mechanisms and functional consequences. Curr Opin Neurol 14:271-278.

Bjartmar C, Wujek JR, Trapp BD (2003) Axonal loss in the pathology of MS: consequences for understanding the progressive phase of the disease. J Neurol Sci 206:165-171.

Bolanos JP, Garcia-Nogales P, Almeida A (2004) Provoking neuroprotection by peroxynitrite. Curr Pharm Des 10:867-877.

Campana WM, Myers RR (2001) Erythropoietin and erythropoietin receptors in the peripheral nervous system: changes after nerve injury. Faseb J 15:1804-1806.

Campana WM, Myers RR (2003) Exogenous erythropoietin protects against dorsal root ganglion apoptosis and pain following peripheral nerve injury. Eur J Neurosci 18:1497-1506.

Chong ZZ, Kang JQ, Maiese K (2002) Erythropoietin is a novel vascular protectant through activation of Akt1 and mitochondrial modulation of cysteine proteases. Circulation 106:2973-2979.

Coleman MP, Perry VH (2002) Axon pathology in neurological disease: a neglected therapeutic target. Trends Neurosci 25:532-537.

Crofton KM, Padilla S, Tilson HA, Anthony DC, Raymer JH, MacPhail RC (1996) The impact of dose rate on the neurotoxicity of acrylamide: the interaction of administered dose, target tissue concentrations, tissue damage, and functional effects. Toxicol Appl Pharmacol 139:163-176.

DeGrandchamp RL, Lowndes HE (1990) Early degeneration and sprouting at the rat neuromuscular junction following acrylamide administration. Neuropathol Appl Neurobiol 16:239-254.

Digicaylioglu M, Lipton SA (2001) Erythropoietin-mediated neuroprotection involves cross-talk between Jak2 and NF-kappaB signalling cascades. Nature 412:641-647.

Ehlers MD (2004) Deconstructing the axon: Wallerian degeneration and the ubiquitin-proteasome system. Trends Neurosci 27:3-6.

Fullerton PM, Barnes JM (1966) Peripheral neuropathy in rats produced by acrylamide. Br J Ind Med 23:210-221.

Glass JD, Culver DG, Levey AI, Nash NR (2002) Very early activation of m-calpain in peripheral nerve during Wallerian degeneration. J Neurol Sci 196:9-20.

Gold BG, Griffin JW, Price DL (1985) Slow axonal transport in acrylamide neuropathy: different abnormalities produced by single-dose and continuous administration. J Neurosci 5:1755-1768.

Gorio A, Gokmen N, Erbayraktar S, Yilmaz O, Madaschi L, Cichetti C, Di Giulio AM, Vardar E, Cerami A, Brines M (2002) Recombinant human erythropoietin counteracts secondary injury and markedly enhances neurological recovery from experimental spinal cord trauma. Proc Natl Acad Sci U S A 99:9450-9455.

Haga KK, Gregory LJ, Hicks CA, Ward MA, Beech JS, Bath PW, Williams SC, O'Neill MJ (2003) The neuronal nitric oxide synthase inhibitor, TRIM, as a neuroprotective agent: effects in models of cerebral ischaemia using histological and magnetic resonance imaging techniques. Brain Res 993:42-53.

Keilhoff G, Fansa H, Wolf G (2002) Differences in peripheral nerve degeneration/regeneration between wild-type and neuronal nitric oxide synthase knockout mice. J Neurosci Res 68:432-441.

Keswani SC, Pardo CA, Cherry CL, Hoke A, McArthur JC (2002) HIV-associated sensory neuropathies. Aids 16:2105-2117.

Keswani SC, Chander B, Hasan C, Griffin JW, McArthur JC, Hoke A (2003a) FK506 is neuroprotective in a model of antiretroviral toxic neuropathy. Ann Neurol 53:57-64.

Keswani SC, Polley M, Pardo CA, Griffin JW, McArthur JC, Hoke A (2003b) Schwann cell chemokine receptors mediate HIV-1 gp120 toxicity to sensory neurons. Ann Neurol 54:287-296.

Keswani SC, Buldanlioglu U, Fischer A, Reed N, Polley M, Liang H, Zhou C, Jack C, Leitz GJ, Hoke A (2004) A novel endogenous erythropoietin mediated pathway prevents axonal degeneration. Ann Neurol 56:815-826.

Ko MH, Chen WP, Lin-Shiau SY, Hsieh ST (1999) Age-dependent acrylamide neurotoxicity in mice: morphology, physiology, and function. Exp Neurol 158:37-46.

Korhonen L, Lindholm D (2004) The ubiquitin proteasome system in synaptic and axonal degeneration: a new twist to an old cycle. J Cell Biol 165:27-30.

LoPachin RM, Jr., Lehning EJ (1994) Acrylamide-induced distal axon degeneration: a proposed mechanism of action. Neurotoxicology 15:247-259.

Mizusawa I, Abe S, Kanno K, Yabashi A, Honda T, Suto M, Hiraiwa K (2003) Expression of cytokines, neurotrophins, neurotrophin receptors and NOS mRNA in dorsal root ganglion of a rat tourniquet model. Leg Med (Tokyo) 5 Suppl 1:S271-274.

Pardo CA, McArthur JC, Griffin JW (2001) HIV neuropathy: insights in the pathology of HIV peripheral nerve disease. J Peripher Nerv Syst 6:21-27.

Raff MC, Whitmore AV, Finn JT (2002) Axonal self-destruction and neurodegeneration. Science 296:868-871.

Ruscher K, Freyer D, Karsch M, Isaev N, Megow D, Sawitzki B, Priller J, Dirnagl U, Meisel A (2002) Erythropoietin is a paracrine mediator of ischemic tolerance in the brain: evidence from an in vitro model. J Neurosci 22:10291-10301.

Sidenius P (1982) The axonopathy of diabetic neuropathy. Diabetes 31:356-363.

Siren AL, Fratelli M, Brines M, Goemans C, Casagrande S, Lewczuk P, Keenan S, Gleiter C, Pasquali C, Capobianco A, Mennini T, Heumann R, Cerami A, Ehrenreich H, Ghezzi P (2001) Erythropoietin prevents neuronal apoptosis after cerebral ischemia and metabolic stress. Proc Natl Acad Sci U S A 98:4044-4049.

Wiggins AK, Shen PJ, Gundlach AL (2003) Neuronal-NOS adaptor protein expression after spreading depression: implications for NO production and ischemic tolerance. J Neurochem 87:1368-1380.

Zhai Q, Wang J, Kim A, Liu Q, Watts R, Hoopfer E, Mitchison T, Luo L, He Z (2003) Involvement of the ubiquitin-proteasome system in the early stages of wallerian degeneration. Neuron 39:217-225.

Chapter 12

ROLE OF ERYTHROPOIETIN IN INFLAMMATORY PATHOLOGIES OF THE CNS

Pietro Ghezzi, Paolo Bigini, Manuela Mengozzi
"Mario Negri" Institute, 20157 Milan, Italy

Abstract: Erythropoietin (EPO) protects against ischemic and traumatic injury of the central nervous system (CNS). In these pathologies, neuronal death is accompanied by an inflammatory reaction including production of inflammatory cytokines, leukocyte infiltration, and astroglial activation/ proliferation. In models of middle cerebral artery occlusion, EPO decreases production of tumor necrosis factor (TNF), interleukin (IL)-6, and of monocyte chemoattractant protein-1 (MCP-1). EPO is also protective in experimental autoimmune encephalomyelitis (EAE) in rats and mice, a disease with a strong inflammatory component where EPO decreases cytokine production and inflammation. Thus, EPO negatively regulates production of inflammatory cytokines in the CNS and, ultimately, acts as an anti-neuroinflammatory agent. On the other hand, inflammatory cytokines IL-1 and TNF negatively regulate EPO production indicating that one of the mechanisms by which inflammation promotes and extend damage could be through inhibition of EPO production. Thus, EPO should be viewed as part of the inflammation/anti-inflammation network in the CNS.

Key words: inflammation, interleukins, tumor necrosis factor, experimental autoimmune encephalomyelitis, multiple sclerosis, glia.

1. INFLAMMATORY PATHOLOGIES OF THE CNS

Many chapters in this book describe a protective effect of EPO in several models of diseases of the nervous system (a list of disease models where EPO offers protection is given in a recent review (Ghezzi and Brines, 2004)). Inflammation is implicated in the pathogenesis of all these diseases, thus a study of the mechanism of the neuroprotective action of EPO must also include a characterization of the effect of EPO on neuroinflammation.

This chapter will deal with the following topics: a description of the anti-inflammatory effects of EPO in stroke; the efficacy of EPO in a neuroinflammatory disease, EAE; the possible mechanisms underlying EPO's anti-neuroinflammatory effects, with a particular attention to its effect on cytokine production. Finally, we will outline a pathway by which inflammatory cytokines can inhibit EPO production, which might explain why endogenous EPO may not be produced at sufficient amounts in CNS diseases with an inflammatory component.

The evidence of an anti-neuroinflammatory action of EPO was clear from our earlier studies on the therapeutic effect of systemically-administered EPO in blunt trauma showing that EPO reduces the inflammatory response and monocytic infiltrate (Brines et al., 2000).

Inflammation is one of the hallmarks of ischemic and traumatic injury of the brain and the spinal cord. In cerebral ischemia, inflammation involves recruitment and influx of vascular leukocytes into the injured brain, activation of resident brain cells and expression of proinflammatory cytokines and adhesion molecules (Dirnagl et al. 1999). Inflammatory cells can aggravate and contribute to brain injury by causing barrier damage, microvascular occlusion and production of a variety of mediators toxic under certain circumstances, including cytokines, reactive oxygen and nitrogen metabolites, and lipid mediators (Witko-Sarsat et al. 2000). The importance of the inflammatory component in traumatic and ischemic brain or spinal cord injury is strengthened by the protective effects of anti-cytokine antibodies or cytokine antagonist or knockouts in many of the animal models.

While the most popular aspect of the role of cytokines in CNS diseases is a pathogenic one, distinctions must be made between acute models of injury (such as ischemia or trauma), chronic neurodegenerative diseases (such as Alzheimer, Parkinson, diabetic neuropathy) and autoimmune diseases (multiple sclerosis), as well as between direct and indirect effects of cytokines. In fact, there are two ways by which cytokines can contribute to neuronal damage. One, indirect, is by activating resident glial or astrocytic cells or inducing recruitment of leukocytes; consequently inflammatory cells will produce toxic species. However, some cytokines can be directly toxic to neurons, as it was reported for interleukin-1 (IL-1) and tumor necrosis factor (TNF).

Although most of the literature in the field investigates or describes a pathogenic role of inflammatory cytokines, many like to view inflammation, with an evolutionary and teleological perspective, as a protective response of the host, by analogy with infectious diseases. In this respect, several cytokines of the IL-6 family (IL-6, IL-11, ciliary neurotrophic factor, leukemia inhibitory factor; oncostatin M, novel neurotrophin-1 and neuropoietin), including IL-6 itself, often regarded mainly as an inflammatory mediator, have neurotrophic activities (Ikeda et al., 1996; Ali et al., 2000; Pizzi et al., 2004). Also TNF, through the receptor TNFR2, may have neuroprotective activities (Shen et al., 1997). This should be kept in mind when interpreting the existing literature of cytokines in CNS diseases.

2. CEREBRAL ISCHEMIA AS AN INFLAMMATORY DISEASE OF THE CNS

Originally, the neuroprotective effect of EPO was investigated in a model of hypoxic-ischemic injury. In fact, adaptation to hypoxia involves not only, at the physiological levels, the increase in red blood cell mass to allow better tissue oxygenation but also cytoprotection, particularly in neurons that are exquisitely sensitive to hypoxic damage. It was interesting to observe that EPO might mediate both physiological adaptation and direct neuroprotection.

Cerebral ischemia is one of the most studied CNS diseases in terms of role of inflammation as a pharmacological target. Earlier studies, including some from our group (Meistrell et al., 1997), have shown that inhibition of TNF using anti-TNF antibodies have protective effect in a rat model where cerebral ischemia is induced by occlusion of the middle cerebral artery (MCA). From then on, several reports have shown that inhibition of specific inflammatory cytokines has protective effects in cerebral ischemia. Inhibition of IL-1 has generally protective effects, as shown by either administration of IL-1ra, knocking out IL-1 genes (Boutin et al., 2001) or inhibiting IL-1 processing with caspase 1 inhibitors (Rabuffetti et al., 2000). Inhibition of fraktalkine by gene disruption (Soriano et al., 2002), and of IL-8 by an antibody (Matsumoto et al., 1997) are also protective, indicating the importance of chemokines in this pathology. On the other hand inhibition of IL-6 by gene disruption has no effect (Clark et al., 2000). On the contrary, mice deficient for the anti-inflammatory cytokine IL-10 show larger infarcts in a model of MCA occlusion (MCAO) (Grilli et al., 2000).

With this background, we undertook a study to specifically investigate the effect of EPO on the inflammatory component of cerebral ischemia using a model of MCAO. The model had been originally used to study the neuroprotective action of EPO (Brines et al., 2000), and is often referred to as "three vessels occlusion". The model consists of complete occlusion of the right MCA and carotid artery, and a reversible, 1-h occlusion of the contralateral carotid artery. This produces a large penumbral region of ischemia in the right frontal cortex.

In this model, administration of EPO markedly diminishes neuronal apoptosis as detected by Tunel stain (Siren et al., 2001). On the inflammatory side, this model produces marked glial and astrocytic activation and inflammatory infiltrate, as detected by immunohistochemistry using anti-CD11b and anti-glial fibrillary acidic protein (GFAP) antibodies, as well as local production of TNF, IL-6 and of the chemokine MCP-1 (Villa et al., 2003). Administration of EPO reduces all these inflammatory markers and mediators, thus suggesting that anti-inflammation may have a role in its protective effect. A representative picture of the anti-inflammatory action of EPO in MCAO, using anti- CD11b antibodies, is shown in Fig.12-1, and is similar to what published by Villa et al. (Villa et al., 2003).

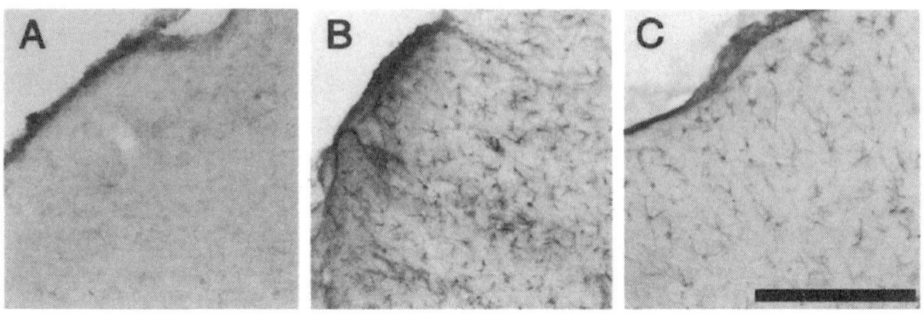

Figure 12-1. Anti-inflammatory effects of EPO in experimental stroke. Representative microphotographs of frontal cortical sections adjacent to the region of necrosis stained with CD11b antibody. They correspond respectively to: sham operated rat (A), ischemic rat treated with PBS (B), and ischemic rat treated with rhEPO (C). As expected, in sham operated rats no CD11b immunopositivity was found. In the cortical regions from vehicle treated ischemic rats an evident pattern of microglial activation, with a great number of CD11b positive cells with thick ramifications and a visible cell body, was observed. The CD11b staining was significantly reduced in ischemic rat treated with rhEPO compared with vehicle treated rats. It is in fact possible to observe as, in the rhEPO treated rats, microglial cells resulted thinner, more ramified and reduced in number respect to vehicle treated rats. Bar, 60 μm.

It is important to note that the inhibition of cytokine production observed in MCAO is clearly independent of the erythropoietic effect of EPO, as it was observed with a single EPO injection, insufficient to increase the hematocrit (this requires approximately a week of daily injections) (Villa et al., 2003). Also, inhibition of inflammatory cytokine production in MCAO was also observed with carbamylated EPO, a non-erythropoietic derivative of EPO that does not bind the classical EPO receptor (Leist et al., 2004).

3. EFFECT OF EPO IN EXPERIMENTAL AUTOIMMUNE ENCEPHALOMYELITIS

The observed marked anti-inflammatory action of EPO in cerebral ischemia prompted us to study its effect on another, completely different, model of an inflammatory disease of the CNS, experimental autoimmune encephalomyelitis (EAE), which is considered the animal model for multiple sclerosis (MS). A variety of EAE models exist, in different animal species, and inhibition of inflammatory cytokines, including knock out mice, antibodies and various inhibitors, have protective effects (stemming from the original report of protective effects on anti-TNF by Ruddle and colleagues (Ruddle et al., 1990)).

We initially used a rat model where acute disease is actively induced in Lewis rats by immunization with myelin basic protein in Freund's adjuvant. This induces an acute disease with a peak at 12-13 days after immunization that recovers by day 20. In this model, a strong inflammatory component is observed, which includes induction of inflammatory cytokines, leukocytic infiltration and astroglial activation. Using this model, preventive administration of EPO from day 3 after immunization (three weekly, at the dose of 50 μg/kg) delayed the onset of the disease and decreased its severity (Agnello et al., 2002). This was paralleled by a marked reduction in the inflammation in the spinal cord (CD11b, GFAP) and in reduced spinal cord levels of inflammatory cytokines TNF and IL-6 (Agnello et al., 2002).

On the other hand, the acute model, with disease signs lasting only a few days, did not allow us to investigate whether EPO can improve EAE when administered in a therapeutic, not preventive, fashion, i.e. after the onset of symptoms. Thus, we switched to a chronic EAE model in C57BL/6 mice immunized with MOG35-55 peptide. Using this model, we could show that EPO, administered several weeks after the appearance of clinical signs of EAE, had a therapeutic effect and diminished the inflammatory response in the spinal cord (P. Ghezzi, manuscript in preparation; and Leist et al., 2004). It is important to note that in both models the effect was independent on the increase in hematocrit as non-erythropoietic carbamylated EPO (Leist et al., 2004), was also effective.

We think that the effectiveness of EPO in EAE may be largely due to its anti-neuroinflammatory effects described above. Although in our hands, inflammation (in terms of cytokine levels, GFAP and CD11b immunostaining) is far less marked in the chronic mouse model compared to the acute rat model, it is well known that this model is responsive to pharmacological strategies targeting inflammation.

Other mechanisms, in addition to anti-inflammation, may thus be involved in the protective effect of EPO in EAE. The anti-apoptotic action of EPO, thought to play a key role in its efficacy in models of neuronal death and neurodegeneration, may not offer a clear hypothesis for its activity in EAE. On one hand, apoptosis of oligodendrocytes has been reported in multiple sclerosis (Dowling et al., 1997; Bitsch et al., 2000; Hisahara et al., 2001; Hisahara et al., 2003), and inhibition of oligodendrocyte apoptosis by baculovirus p35 caspase inhibitor protects from EAE (Hisahara et al., 2000). However, the role of oligodendrocyte apoptosis in MS and EAE is far from being univocally defined, and there are reports indicating that effector T cells, rather than the target oligodendrocytes, undergo apoptosis (Bonetti et al., 1997). The latter observation is probably more in agreement with the most accepted view that apoptosis (namely, of T cells) has a protective effect in EAE (Critchfield and Lenardo, 1995; White et al., 1998; Pender, 1999; Pender and Rist, 2001) and in vivo administration of apoptosis inhibitors such as zVAD-fmk (Okuda et al., 2000) worsen EAE, probably by inhibition of lymphocyte apoptosis and suppressed apoptotic death of inflammatory cells. Intrathecally-administered interferon-gamma also has a protective effect on a chronic model of EAE in mice by increasing apoptosis in T cells infiltrating the CNS (Furlan et al., 2001), and uptake of apoptotic T cells by microglia is tolerogenic and results in a reduced autoimmune inflammation (Magnus et al., 2001).

Additional mechanisms could also contribute to the effect of EPO in EAE. For instance, excitotoxicity plays a role in EAE, as shown by protective effect of AMPA or kainate antagonists (Pitt et al., 2000; Smith et al., 2000). This excitotoxic pathway may be a common pathogenic mechanism for EAE (Pitt et al., 2000) and stroke (Dirnagl et al., 1999). Interestingly our earlier report indicate a protective effect of EPO on kainate toxicity (Brines et al., 2000).

In conclusion, additional studies need to be performed to clarify the mechanism of action of EPO in EAE. The pathogenesis of this autoimmune disease is clearly more complicated than that of ischemic injury. In this respect, it is important to note that EPO may act on mechanisms that are common to both EAE and ischemia (such as excitotoxicity as mentioned above. Like in EAE, oligodendrocytes are also a primary target in cerebral ischemia (Dewar et al., 2003). From a more general point of view of the physiological response to injury, it is interesting to note that several recent papers have indicated the possibility of hypoxic-like damage in multiple sclerosis, mostly suggested by the activation of hypoxia-responsive genes, including induction of several HIF target genes (such as VEGF, but not EPO, it should be noted) (Proescholdt et al., 2002; Aboul-Enein et al., 2003; Graumann et al., 2003; Kirk and Karlik, 2003; Lassmann, 2003).

4. ANTI-INFLAMMATORY ACTION OF EPO: MECHANISM OF ACTION

As discussed above, EPO decreases inflammation in traumatic brain injury, cerebral ischemia and EAE. We also observed a diminished inflammation in other models, including diabetic neuropathy (unpublished data), where EPO has a neuroprotective action (Bianchi et al., 2004). Is then EPO an anti-inflammatory cytokine? The term "anti-inflammatory cytokine" was originally used for IL-4 and IL-10 and later extended to IL-13 and TGF-beta as these cytokines inhibit the synthesis of inflammatory cytokines, including TNF, by monocytes in vitro and/or in other experimental models. Having this in mind, and given the inhibition of the local production of TNF in the experimental models described above, we investigated the effect of EPO on cytokine production in vitro using human monocytes or rat glial cells. In these experiments, TNF production was induced with a classical stimulus (lipopolysaccharide, LPS). To our surprise, we could not observe any inhibition with EPO at any of the concentrations tested (Villa et al., 2003). This lack of "anti-inflammatory" effect of EPO was also confirmed in an in vivo model where local TNF production is induced in the brain by an intracerebroventricular injection of LPS. In this model, where IL-10 and IL-13 markedly inhibit TNF production, EPO is without an effect (Villa et al., 2003). Squadrito et al. also tested the effect of EPO on serum TNF levels induced by splanchnic artery occlusion and found that, while EPO had protective effects in this disease model, TNF production was unaffected (Squadrito et al., 1999). It should be noted, however, that EPO was reported to inhibit the induction of NO production by LPS in normal rat peritoneal macrophages in vitro, suggesting that EPO may have some direct "anti-inflammatory" actions (Squadrito et al., 1999).

The wide use of EPO in hemodialysis patients has allowed a few ex vivo studies on inflammatory responses in these patients. A previous study showed that whole blood obtained from hemodialysis patients treated with EPO for 6 months causes an increase in IL-10 production and transiently decreases TNF, suggesting that EPO treatment might reduce the inflammatory process in these patients (Bryl et al., 1998). Buemi et al. reported that, in hemodialyzed subjects, EPO decreased the production of the marker of macrophage activation, neopterin (Buemi et al., 1991). However, in another report, the chemotactic response of polymorphonuclear leukocytes was unaffected (Sperschneider et al., 1996). It should, however, be noted that the basal plasma levels of IL-1alpha in hemodialysis patients are slightly elevated by EPO treatment, although the significance of baseline IL-1 is questionable, particularly in terms of possible role in inflammation.

To explain the observed inhibition of inflammation (including inhibition of cytokine production) by EPO in cerebral ischemia we propose a model where EPO acts primarily through an anti-apoptotic mechanism. While apoptosis is commonly viewed as a way of dying without inducing inflammation, there are examples that apoptosis, in some contexts, may promote inflammation. For instance, in cerebral ischemia (Rabuffetti et al., 2000) and in a model of renal ischemia (de Vries et al., 2003), inhibition of apoptosis by lysophosphatidic acid or caspase inhibitors reduce the inflammatory response, and production of inflammatory cytokines by tissue macrophages during phagocytosis of apoptotic cells (Kurosaka et al., 2001; Canbay et al., 2003).

To test this hypothesis, we used a model of mixed cultures of glial cells and hippocampal neurons where neuronal death is induced by the selective neurotoxic agent trimethyl tin (TMT). In this model, increased glial TNF production and release is a strict function of neuronal death from apoptosis (Viviani et al., 1998; Viviani et al., 2000). When tested in this system, EPO decreased TMT-induced death in hippocampal neurons and the associated production of TNF by the glial cells in co-culture (Villa et al., 2003). This would suggest that EPO exerts an anti-inflammatory effect only in settings where inflammation is secondary to neuronal death. A similar mechanism has been proposed by Chong et al. (Chong et al., 2003). They have demonstrated that addition of EPO to primary hippocampal neurons in vitro prevents apoptosis and externalization of phosphatidyl serine (PS) residues (Chong et al., 2003). Since it was known in the literature that PS exposure by neurons induces their phagocytosis by microglial cells (Maiese and Vincent, 2000; Hoffmann et al., 2001), they investigated the effect of EPO and reported that it inhibits microglial activation and microglial PS receptor expression, and both effects ultimately reduced microglial phagocytosis of neurons (Chong et al., 2003). This possible mechanism is outlined in Fig. 12-2.

In addition to this mechanism where neuronal death induces inflammation, hypoxia augments production of inflammatory cytokines, an effect described first by us for IL-1 and TNF in 1991 (Ghezzi et al., 1991), and later observed for other cytokines including chemokines (Karakurum et al., 1994). It is not known, to date, whether EPO influences hypoxia-induced cytokines.

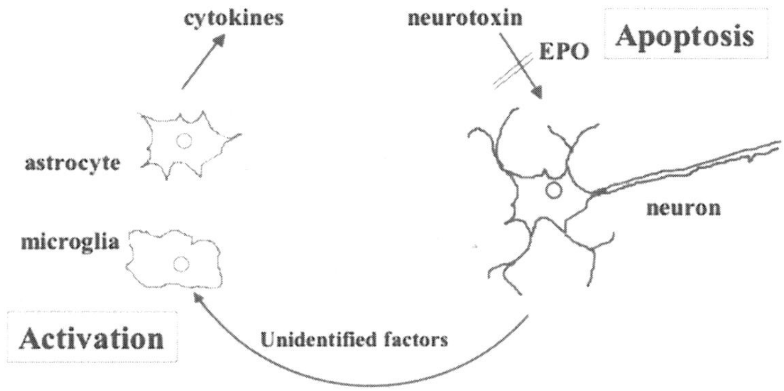

Figure 12-2. Interplay between the anti-apoptotic action of EPO and neuroinflammation in a context where inflammation is initiated by products of dying neurons.

5. ANTI-INFLAMMATORY ACTION OF EPO IN EAE: EVIDENCE FOR ADDITIONAL MECHANISM ACTION?

The reduced inflammation observed in this EAE suggests that mechanisms different from the one outlined above, mediated by the anti-apoptotic action of EPO, might be involved. Although oligodendrocyte apoptosis occurs in EAE, it is not thought to be as extensive as the neuronal apoptosis observed in ischemic cerebral cortex. At present, we have no explanation for the anti-inflammatory action of EPO in this context. In fact, we believe that the definition of the mechanism of the anti-inflammatory action of EPO in autoimmunity may point out to additional mechanisms to explain its anti-inflammatory action on cerebral ischemia.

Of the typical signs of inflammation, one of the most studied is the increase in vascular permeability induced by several inflammatory cytokines, which in turn allows infiltration of leukocytes. In the CNS, this is often synonymous with disruption of the blood-brain barrier. In this respect, it is important to note that VEGF, which is also induced as a response to injury and plays a role in neuro-repair, increases vascular permeability (Dvorak et al., 1995). This effect is also observed using in vitro models of the blood-brain barrier function (Martinez-Estrada et al., 2003). EPO, which shares angiogenic and neurotrophic/neuroprotective actions with VEGF (Sun et al., 2003), does not have such an effect, but it counteracts the VEGF-induced permeability (Martinez-Estrada et al., 2003). Since VEGF is induced in EAE and MS (Proescholdt et al., 2002; Aboul-Enein et al., 2003; Graumann et al., 2003; Kirk and Karlik, 2003; Lassmann, 2003), as well as in cerebral ischemia (Hayashi et al., 1997), and may have pro-neuroinflammatory activities when locally applied (Proescholdt et al., 2002), the inhibition of VEGF actions by EPO might contribute to the observed reduction of the inflammatory infiltrate observed in these diseases.

6. EFFECT OF INFLAMMATORY CYTOKINES AND PROSTAGLANDINS ON EPO PRODUCTION

The role of endogenous EPO in ischemic preconditioning and as a neuroprotective mechanism raises the question of studying the endogenous factors associated with inflammation affecting EPO production. In the CNS so far only one paper showed an inhibitory effect of IL-1 and TNF on EPO production by astrocytes (Nagai et al., 2001). However, there is a consistent literature reporting an inhibitory effect of inflammatory cytokines on EPO synthesis outside the CNS. In fact, the anemia that develops in chronic inflammatory, neoplastic or infectious disorders, known as anemia of chronic disease (ACD), is partly due to impaired erythropoietin synthesis: EPO levels are lower than those observed in a control population with anemia of comparable severity (Hochberg et al., 1988; Miller et al., 1990). A characteristic finding of the disorders associated with ACD is increased production of proinflammatory cytokines, including interleukin-1 (IL-1) and tumor necrosis factor (TNF). Therefore, it has been proposed that proinflammatory cytokines might inhibit hypoxia-induced EPO synthesis.

Since a suitable in vitro model of kidney cells involved in EPO synthesis is still lacking, the effect of proinflammatory cytokines was studied on two human hepatoma cell lines (Hep3B and HepG2) that release EPO in the culture medium when appropriately stimulated.

IL-1 and TNF inhibited EPO mRNA levels and protein secretion in hypoxia-stimulated Hep3B cells (Faquin et al., 1992). Similar results were found by Jelkmann *et al.* (Jelkmann et al., 1992) in HepG2 cells and in isolated perfused rat kidney. When administered in vivo to normoxic rats, LPS inhibited renal EPO mRNA levels, probably through TNF, which increased about ten-fold in rat kidney upon LPS treatment (Frede et al., 1997). In vivo, administration of LPS and IL-1 inhibited hypoxia-induced renal EPO mRNA levels and plasma EPO in rats (Frede et al., 1997).

The most important transcription factor for hypoxic stimulation of EPO synthesis is the hypoxia-inducible factor-1 (HIF-1). However, HIF-1 is not involved in IL-1- and TNF-mediated inhibition of EPO synthesis, since these cytokines increased HIF-1 DNA binding (Hellwig-Burgel et al., 1999). Accordingly, neither IL-1 nor TNF decreased hypoxia-induced VEGF production in HepG2 cells (Hellwig-Burgel et al., 1999). More recent studies have shown that the inhibitory effect of IL-1 and TNF on EPO gene expression is mediated by the transcription factors GATA-2 and NF-kB (La Ferla et al., 2002). The EPO promoter and the 5' flanking region contain binding sites for GATA-2 and NF-kB. In hypoxic conditions, HIF-1 is induced and binds to the EPO 3' enhancer, NF-kB is not affected, and GATA-2 binding is decreased, resulting in increased EPO production (La Ferla et al., 2002). In the presence of IL-1 or TNF, hypoxic induction of HIF-1 is unchanged, and GATA-2 and NF-kB binding are increased, resulting in decreased EPO production (La Ferla et al., 2002) (Fig. 12-3). GATA-2 was already known as an inhibitory factor of EPO gene expression (Ebert and Bunn, 1999), and recently demonstrated to be involved in the repression of EPO by cytokines (La Ferla et al., 2002), whereas NF-kB has been described as a repressor of EPO gene expression only for cytokine-mediated inhibition of EPO (La Ferla et al., 2002).

A

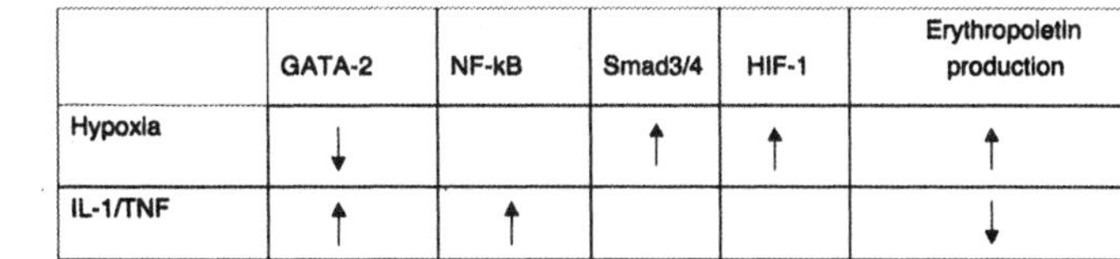

B

	GATA-2	NF-kB	Smad3/4	HIF-1	Erythropoietin production
Hypoxia	↓		↑	↑	↑
IL-1/TNF	↑	↑			↓
TGF-beta			↑		↑

Figure 12-3. Schematic representation of EPO transcriptional regulation upon hypoxia and cytokine treatment. **A**, Relevant binding sites on the 5' promoter and on the 3' enhancer of the erythropoietin gene for transcription factors involved in cytokine regulation and hypoxia induction. HRE and SBE stand for hypoxia responsive element and Smad binding element, respectively. **B**, Schematic drawing of the effect of hypoxia and of cytokine stimulation on the relevant transcription factors for EPO gene regulation.

In contrast to the divergent regulation of EPO and VEGF observed with IL-1 and TNF, interestingly TGF-beta increased both hypoxia-induced EPO and VEGF synthesis (Sanchez-Elsner et al., 2001; Sanchez-Elsner et al., 2004). In this case, increased EPO and VEGF expression by TGF-beta was mediated by Smad proteins, particularly by Smad3/Smad4. The Smad3/Smad4 binding site is located in the 3' enhancer of the EPO gene and in the 5' enhancer of the VEGF genes, where HIF-1 binding sites are also located (Sanchez-Elsner et al., 2001; Sanchez-Elsner et al., 2004).

Concerning the role of prostaglandins (PG), in a series of old studies from Fisher's group (reviewed in ref. (Fisher, 1988)) and Bauer's group (Kurtz et al., 1985), cAMP elevating agents, including prostaglandin PGE_2, adenosine, forskolin, were shown to increase EPO production in perfused dog kidneys, rat mesangial cells, hepatoma cell lines and *in vivo*. Based on the evidence that hypoxia-induced PG synthesis in several models of renal cells in culture (Hagiwara et al., 1984; Jelkmann et al., 1985), and indomethacin, a potent cyclo-oxygenase (COX) inhibitor, inhibited PGE_2 and EPO production in animals exposed to renal artery constriction and hypoxia (Fisher et al., 1978), a role for PGs released within the kidney in the modulation of kidney production of EPO was postulated.

However, the requirement for endogenous PG induction, and cAMP elevation, in hypoxia-induced EPO synthesis is controversial, and depends on the model investigated. In rat mesangial cell cultures, hypoxia increases PG production (Jelkmann et al., 1985), and indomethacin, an inhibitor of COX, and therefore of PG synthesis, blocked hypoxia-induced EPO production (Kurtz et al., 1985). In contrast, in Hep3B cells hypoxia did not increase cAMP accumulation (Nakashima et al., 1992), although the cAMP analogue 8-bromo cAMP, and cAMP elevation induced by forskolin, still increased hypoxia-induced EPO production (Kvietikova et al., 1995).

Whether or not endogenous PGs, and subsequent PG-induced cAMP elevation, are required for hypoxia-induced EPO synthesis, it is clear that cAMP-elevating agents increase hypoxia-induced EPO production. A molecular basis for this effect is the recent demonstration that cAMP analogs increase HIF-1 activity through protein kinase A activation (Kvietikova et al., 1995). Also, a direct effect of PGE_2 on HIF-1 activation, and consequently on VEGF production, has been reported (Liu et al., 2002; Fukuda et al., 2003).

Endogenous PGE_2 production, due to hypoxia-induced COX-2 activation, has recently been shown to be involved in hypoxia-induced VEGF synthesis in several endothelial cell types (Schmedtje et al., 1997; Bonazzi et al., 2000) and in prostate cancer cells (Liu et al., 1999). Accordingly, non-steroidal anti-inflammatory drugs (NSAIDs) inhibit angiogenesis (reviewed in ref. (Romano and Claria, 2003)), probably through inhibition of VEGF (Palayoor et al., 2003). Recently, COX-2 was found to be induced by hypoxia in rat kidney (Ogawa et al., 2002), sustaining the role of endogenous PGs in renal hypoxia-induced EPO, and confirming old data on indomethacin inhibition of *in vivo* hypoxia-induced EPO production (Fisher et al., 1978).

Altogether, these data may suggest a positive role of PGE_2-induced cAMP on EPO production, which is opposite of the inhibitory role of inflammatory cytokines, and is well in agreement with the feedback mechanism by which both phosphodiesterase inhibitors and PGE_2, by elevating cAMP, potently inhibits the synthesis of TNF and other inflammatory cytokines (Spengler et al., 1989; Spinas et al., 1991; Sommer et al., 1995).

These pathways may also explain other, potentially deleterious, effects of inflammation in the context of CNS diseases. For instance, application of exogenous EPO was shown to stimulate neurogenesis (Shingo et al., 2001; Wang et al., 2004), and a recent report has shown that inflammatory activation of microglia can suppress hippocampal neurogenesis in vivo (Ekdahl et al., 2003).

In conclusion, the anti-inflammatory effect of EPO in many models of CNS diseases suggests its possible usefulness in the therapy of neuroinflammatory conditions of diverse origins, and stresses the importance of viewing its role in the context of the classical cytokine network in inflammation, where EPO is normally neglected.

REFERENCES

Aboul-Enein F, Rauschka H, Kornek B, Stadelmann C, Stefferl A, Bruck W, Lucchinetti C, Schmidbauer M, Jellinger K, Lassmann H (2003) Preferential loss of myelin-associated glycoprotein reflects hypoxia-like white matter damage in stroke and inflammatory brain diseases. J Neuropathol Exp Neurol 62:25-33.

Agnello D, Bigini P, Villa P, Mennini T, Cerami A, Brines ML, Ghezzi P (2002) Erythropoietin exerts an anti-inflammatory effect on the CNS in a model of experimental autoimmune encephalomyelitis. Brain Res 952:128-134.

Ali C, Nicole O, Docagne F, Lesne S, MacKenzie ET, Nouvelot A, Buisson A, Vivien D (2000) Ischemia-induced interleukin-6 as a potential endogenous neuroprotective cytokine against NMDA receptor-mediated excitotoxicity in the brain. J Cereb Blood Flow Metab 20:956-966.

Bianchi R, Buyukakilli B, Brines M, Savino C, Cavaletti G, Oggioni N, Lauria G, Borgna M, Lombardi R, Cimen B, Comelekoglu U, Kanik A, Tataroglu C, Cerami A, Ghezzi P (2004) Erythropoietin both protects from and reverses experimental diabetic neuropathy. Proc Natl Acad Sci U S A 101:823-828.

Bitsch A, Kuhlmann T, Da Costa C, Bunkowski S, Polak T, Bruck W (2000) Tumour necrosis factor alpha mRNA expression in early multiple sclerosis lesions: correlation with demyelinating activity and oligodendrocyte pathology. Glia 29:366-375.

Bonazzi A, Mastyugin V, Mieyal PA, Dunn MW, Laniado-Schwartzman M (2000) Regulation of cyclooxygenase-2 by hypoxia and peroxisome proliferators in the corneal epithelium. J Biol Chem 275:2837-2844.

Bonetti B, Pohl J, Gao YL, Raine CS (1997) Cell death during autoimmune demyelination: effector but not target cells are eliminated by apoptosis. J Immunol 159:5733-5741.

Boutin H, LeFeuvre RA, Horai R, Asano M, Iwakura Y, Rothwell NJ (2001) Role of IL-1alpha and IL-1beta in ischemic brain damage. J Neurosci 21:5528-5534.

Brines ML, Ghezzi P, Keenan S, Agnello D, de Lanerolle NC, Cerami C, Itri LM, Cerami A (2000) Erythropoietin crosses the blood-brain barrier to protect against experimental brain injury. Proc Natl Acad Sci U S A 97:10526-10531.

Bryl E, Mysliwska J, Debska-Slizien A, Rachon D, Bullo B, Lizakowski S, Mysliwski A, Rutkowski B (1998) The influence of recombinant human erythropoietin on tumor necrosis factor alpha and interleukin-10 production by whole blood cell cultures in hemodialysis patients. Artif Organs 22:177-181.

Buemi M, Allegra A, Aloisi C, Frisina N (1991) Influence of therapy with recombinant erythropoietin on serum levels of neopterin in hemodialyzed subjects. Am J Nephrol 11:281-283.

Canbay A, Feldstein AE, Higuchi H, Werneburg N, Grambihler A, Bronk SF, Gores GJ (2003) Kupffer cell engulfment of apoptotic bodies stimulates death ligand and cytokine expression. Hepatology 38:1188-1198.

Chong ZZ, Kang JQ, Maiese K (2003) Erythropoietin fosters both intrinsic and extrinsic neuronal protection through modulation of microglia, Akt1, Bad, and caspase-mediated pathways. Br J Pharmacol 138:1107-1118.

Clark WM, Rinker LG, Lessov NS, Hazel K, Hill JK, Stenzel-Poore M, Eckenstein F (2000) Lack of interleukin-6 expression is not protective against focal central nervous system ischemia. Stroke 31:1715-1720.

Critchfield JM, Lenardo MJ (1995) Antigen-induced programmed T cell death as a new approach to immune therapy. Clin Immunol Immunopathol 75:13-19.

de Vries B, Matthijsen RA, van Bijnen AA, Wolfs TG, Buurman WA (2003) Lysophosphatidic acid prevents renal ischemia-reperfusion injury by inhibition of apoptosis and complement activation. Am J Pathol 163:47-56.

Dewar D, Underhill SM, Goldberg MP (2003) Oligodendrocytes and ischemic brain injury. J Cereb Blood Flow Metab 23:263-274.

Dirnagl U, Iadecola C, Moskowitz MA (1999) Pathobiology of ischaemic stroke: an integrated view. Trends Neurosci 22:391-397.

Dowling P, Husar W, Menonna J, Donnenfeld H, Cook S, Sidhu M (1997) Cell death and birth in multiple sclerosis brain. J Neurol Sci 149:1-11.

Dvorak HF, Brown LF, Detmar M, Dvorak AM (1995) Vascular permeability factor/vascular endothelial growth factor, microvascular hyperpermeability, and angiogenesis. Am J Pathol 146:1029-1039.

Ebert BL, Bunn HF (1999) Regulation of the erythropoietin gene. Blood 94:1864-1877.

Ekdahl CT, Claasen JH, Bonde S, Kokaia Z, Lindvall O (2003) Inflammation is detrimental for neurogenesis in adult brain. Proc Natl Acad Sci U S A 100:13632-13637.

Faquin WC, Schneider TJ, Goldberg MA (1992) Effect of inflammatory cytokines on hypoxia-induced erythropoietin production. Blood 79:1987-1994.

Fisher JW (1988) Pharmacologic modulation of erythropoietin production. Annu Rev Pharmacol Toxicol 28:101-122.

Fisher JW, Gross DM, Foley JE, Nelson PK, Rodgers GM, George WJ, Jubiz W (1978) A concept for the control of kidney production of erythropoietin involving prostaglandins and cyclic nucleotides. Contrib Nephrol 13:37-59.

Frede S, Fandrey J, Pagel H, Hellwig T, Jelkmann W (1997) Erythropoietin gene expression is suppressed after lipopolysaccharide or interleukin-1 beta injections in rats. Am J Physiol 273:R1067-1071.

Fukuda R, Kelly B, Semenza GL (2003) Vascular endothelial growth factor gene expression in colon cancer cells exposed to prostaglandin E2 is mediated by hypoxia-inducible factor 1. Cancer Res 63:2330-2334.

Furlan R, Brambilla E, Ruffini F, Poliani PL, Bergami A, Marconi PC, Franciotta DM, Penna G, Comi G, Adorini L, Martino G (2001) Intrathecal delivery of IFN-gamma protects C57BL/6 mice from chronic- progressive experimental autoimmune encephalomyelitis by increasing apoptosis of central nervous system-infiltrating lymphocytes. J Immunol 167:1821-1829.

Ghezzi P, Brines M (2004) Erythropoietin as an antiapoptotic, tissue-protective cytokine. Cell Death Differentiation 11:S37-S44.

Ghezzi P, Dinarello CA, Bianchi M, Rosandich ME, Repine JE, White CW (1991) Hypoxia increases production of interleukin-1 and tumor necrosis factor by human mononuclear cells. Cytokine 3:189-194.

Graumann U, Reynolds R, Steck AJ, Schaeren-Wiemers N (2003) Molecular changes in normal appearing white matter in multiple sclerosis are characteristic of neuroprotective mechanisms against hypoxic insult. Brain Pathol 13:554-573.

Grilli M, Barbieri I, Basudev H, Brusa R, Casati C, Lozza G, Ongini E (2000) Interleukin-10 modulates neuronal threshold of vulnerability to ischaemic damage. Eur J Neurosci 12:2265-2272.

Hagiwara M, McNamara DB, Chen IL, Fisher JW (1984) Role of endogenous prostaglandin E2 in erythropoietin production and dome formation by human renal carcinoma cells in culture. J Clin Invest 74:1252-1261.

Hayashi T, Abe K, Suzuki H, Itoyama Y (1997) Rapid induction of vascular endothelial growth factor gene expression after transient middle cerebral artery occlusion in rats. Stroke 28:2039-2044.

Hellwig-Burgel T, Rutkowski K, Metzen E, Fandrey J, Jelkmann W (1999) Interleukin-1beta and tumor necrosis factor-alpha stimulate DNA binding of hypoxia-inducible factor-1. Blood 94:1561-1567.

Hisahara S, Okano H, Miura M (2003) Caspase-mediated oligodendrocyte cell death in the pathogenesis of autoimmune demyelination. Neurosci Res 46:387-397.

Hisahara S, Yuan J, Momoi T, Okano H, Miura M (2001) Caspase-11 mediates oligodendrocyte cell death and pathogenesis of autoimmune-mediated demyelination. J Exp Med 193:111-122.

Hisahara S, Araki T, Sugiyama F, Yagami K, Suzuki M, Abe K, Yamamura K, Miyazaki J, Momoi T, Saruta T, Bernard CC, Okano H, Miura M (2000) Targeted expression of baculovirus p35 caspase inhibitor in oligodendrocytes protects mice against autoimmune-mediated demyelination. Embo J 19:341-348.

Hochberg MC, Arnold CM, Hogans BB, Spivak JL (1988) Serum immunoreactive erythropoietin in rheumatoid arthritis: impaired response to anemia. Arthritis Rheum 31:1318-1321.

Hoffmann PR, deCathelineau AM, Ogden CA, Leverrier Y, Bratton DL, Daleke DL, Ridley AJ, Fadok VA, Henson PM (2001) Phosphatidylserine (PS) induces PS receptor-mediated macropinocytosis and promotes clearance of apoptotic cells. J Cell Biol 155:649-659.

Ikeda K, Kinoshita M, Tagaya N, Shiojima T, Taga T, Yasukawa K, Suzuki H, Okano A (1996) Coadministration of interleukin-6 (IL-6) and soluble IL-6 receptor delays progression of wobbler mouse motor neuron disease. Brain Res 726:91-97.

Jelkmann W, Pagel H, Wolff M, Fandrey J (1992) Monokines inhibiting erythropoietin production in human hepatoma cultures and in isolated perfused rat kidneys. Life Sci 50:301-308.

Jelkmann W, Kurtz A, Forstermann U, Pfeilschifter J, Bauer C (1985) Hypoxia enhances prostaglandin synthesis in renal mesangial cell cultures. Prostaglandins 30:109-118.

Karakurum M, Shreeniwas R, Chen J, Pinsky D, Yan SD, Anderson M, Sunouchi K, Major J, Hamilton T, Kuwabara K, et al. (1994) Hypoxic induction of interleukin-8 gene expression in human endothelial cells. J Clin Invest 93:1564-1570.

Kirk SL, Karlik SJ (2003) VEGF and vascular changes in chronic neuroinflammation. J Autoimmun 21:353-363.

Kurosaka K, Watanabe N, Kobayashi Y (2001) Production of proinflammatory cytokines by resident tissue macrophages after phagocytosis of apoptotic cells. Cell Immunol 211:1-7.

Kurtz A, Jelkmann W, Pfeilschifter J, Bauer C (1985) Role of prostaglandins in hypoxia-stimulated erythropoietin production. Am J Physiol 249:C3-8.

Kvietikova I, Wenger RH, Marti HH, Gassmann M (1995) The transcription factors ATF-1 and CREB-1 bind constitutively to the hypoxia-inducible factor-1 (HIF-1) DNA recognition site. Nucleic Acids Res 23:4542-4550.

La Ferla K, Reimann C, Jelkmann W, Hellwig-Burgel T (2002) Inhibition of erythropoietin gene expression signaling involves the transcription factors GATA-2 and NF-kappaB. Faseb J 16:1811-1813.

Lassmann H (2003) Hypoxia-like tissue injury as a component of multiple sclerosis lesions. J Neurol Sci 206:187-191.

Leist M, Ghezzi P, Grasso G, Bianchi R, Villa P, Fratelli M, Savino C, Bianchi M, Nielsen J, Gerwien J, Kallunki P, Larsen AK, Helboe L, Christensen S, Pedersen LO, Nielsen M, Torup L, Sager T, Sfacteria A, Erbayraktar S, Erbayraktar Z, Gokmen N, Yilmaz O, Cerami-Hand C, Xie Q-w, Coleman T, Cerami A, Brines M (2004) Derivatives of Erythropoietin That Are Tissue Protective But Not Erythropoietic. Science 305:239-242.

Liu XH, Kirschenbaum A, Yao S, Stearns ME, Holland JF, Claffey K, Levine AC (1999) Upregulation of vascular endothelial growth factor by cobalt chloride-simulated hypoxia is mediated by persistent induction of cyclooxygenase-2 in a metastatic human prostate cancer cell line. Clin Exp Metastasis 17:687-694.

Liu XH, Kirschenbaum A, Lu M, Yao S, Dosoretz A, Holland JF, Levine AC (2002) Prostaglandin E2 induces hypoxia-inducible factor-1alpha stabilization and nuclear localization in a human prostate cancer cell line. J Biol Chem 277:50081-50086.

Magnus T, Chan A, Grauer O, Toyka KV, Gold R (2001) Microglial phagocytosis of apoptotic inflammatory T cells leads to down- regulation of microglial immune activation. J Immunol 167:5004-5010.

Maiese K, Vincent AM (2000) Membrane asymmetry and DNA degradation: functionally distinct determinants of neuronal programmed cell death. J Neurosci Res 59:568-580.

Martinez-Estrada OM, Rodriguez-Millan E, Gonzalez-De Vicente E, Reina M, Vilaro S, Fabre M (2003) Erythropoietin protects the in vitro blood-brain barrier against VEGF-induced permeability. Eur J Neurosci 18:2538-2544.

Matsumoto T, Ikeda K, Mukaida N, Harada A, Matsumoto Y, Yamashita J, Matsushima K (1997) Prevention of cerebral edema and infarct in cerebral reperfusion injury by an antibody to interleukin-8. Lab Invest 77:119-125.

Meistrell ME, 3rd, Botchkina GI, Wang H, Di Santo E, Cockroft KM, Bloom O, Vishnubhakat JM, Ghezzi P, Tracey KJ (1997) Tumor necrosis factor is a brain damaging cytokine in cerebral ischemia. Shock 8:341-348.

Miller CB, Jones RJ, Piantadosi S, Abeloff MD, Spivak JL (1990) Decreased erythropoietin response in patients with the anemia of cancer. N Engl J Med 322:1689-1692.

Nagai A, Nakagawa E, Choi HB, Hatori K, Kobayashi S, Kim SU (2001) Erythropoietin and erythropoietin receptors in human CNS neurons, astrocytes, microglia, and oligodendrocytes grown in culture. J Neuropathol Exp Neurol 60:386-392.

Nakashima J, Brookins J, Fisher JW (1992) Characterization of erythropoietin production in a hepatocellular carcinoma cell line. J Lab Clin Med 119:306-314.

Ogawa T, Tomomasa T, Morikawa A (2002) Developmental changes in cyclo-oxygenase mRNA induction by hypoxia in rat kidney. Pediatr Int 44:675-679.

Okuda Y, Sakoda S, Fujimura H, Yanagihara T (2000) The effect of apoptosis inhibitors on experimental autoimmune encephalomyelitis: apoptosis as a regulatory factor. Biochem Biophys Res Commun 267:826-830.

Palayoor ST, Tofilon PJ, Coleman CN (2003) Ibuprofen-mediated reduction of hypoxia-inducible factors HIF-1alpha and HIF-2alpha in prostate cancer cells. Clin Cancer Res 9:3150-3157.

Pender MP (1999) Activation-induced apoptosis of autoreactive and alloreactive T lymphocytes in the target organ as a major mechanism of tolerance. Immunol Cell Biol 77:216-223.

Pender MP, Rist MJ (2001) Apoptosis of inflammatory cells in immune control of the nervous system: role of glia. Glia 36:137-144.

Pitt D, Werner P, Raine CS (2000) Glutamate excitotoxicity in a model of multiple sclerosis. Nat Med 6:67-70.

Pizzi M, Sarnico I, Boroni F, Benarese M, Dreano M, Garotta G, Valerio A, Spano P (2004) Prevention of neuron and oligodendrocyte degeneration by interleukin-6 (IL-6) and IL-6 receptor/IL-6 fusion protein in organotypic hippocampal slices. Mol Cell Neurosci 25:301-311.

Proescholdt MA, Jacobson S, Tresser N, Oldfield EH, Merrill MJ (2002) Vascular endothelial growth factor is expressed in multiple sclerosis plaques and can induce inflammatory lesions in experimental allergic encephalomyelitis rats. J Neuropathol Exp Neurol 61:914-925.

Rabuffetti M, Sciorati C, Tarozzo G, Clementi E, Manfredi AA, Beltramo M (2000) Inhibition of caspase-1-like activity by Ac-Tyr-Val-Ala-Asp-chloromethyl ketone induces long-lasting neuroprotection in cerebral ischemia through apoptosis reduction and decrease of proinflammatory cytokines. J Neurosci 20:4398-4404.

Romano M, Claria J (2003) Cyclooxygenase-2 and 5-lipoxygenase converging functions on cell proliferation and tumor angiogenesis: implications for cancer therapy. Faseb J 17:1986-1995.

Ruddle NH, Bergman CM, McGrath KM, Lingenheld EG, Grunnet ML, Padula SJ, Clark RB (1990) An antibody to lymphotoxin and tumor necrosis factor prevents transfer of experimental allergic encephalomyelitis. J Exp Med 172:1193-1200.

Sanchez-Elsner T, Botella LM, Velasco B, Corbi A, Attisano L, Bernabeu C (2001) Synergistic cooperation between hypoxia and transforming growth factor-beta pathways on human vascular endothelial growth factor gene expression. J Biol Chem 276:38527-38535.

Sanchez-Elsner T, Ramirez JR, Sanz-Rodriguez F, Varela E, Bernabeu C, Botella LM (2004) A cross-talk between hypoxia and TGF-beta orchestrates erythropoietin gene regulation through SP1 and Smads. J Mol Biol 336:9-24.

Schmedtje JF, Jr., Ji YS, Liu WL, DuBois RN, Runge MS (1997) Hypoxia induces cyclooxygenase-2 via the NF-kappaB p65 transcription factor in human vascular endothelial cells. J Biol Chem 272:601-608.

Shen Y, Li R, Shiosaki K (1997) Inhibition of p75 tumor necrosis factor receptor by antisense oligonucleotides increases hypoxic injury and beta-amyloid toxicity in human neuronal cell line. J Biol Chem 272:3550-3553.

Shingo T, Sorokan ST, Shimazaki T, Weiss S (2001) Erythropoietin regulates the in vitro and in vivo production of neuronal progenitors by mammalian forebrain neural stem cells. J Neurosci 21:9733-9743.

Siren AL, Fratelli M, Brines M, Goemans C, Casagrande S, Lewczuk P, Keenan S, Gleiter C, Pasquali C, Capobianco A, Mennini T, Heumann R, Cerami A, Ehrenreich H, Ghezzi P (2001) Erythropoietin prevents neuronal apoptosis after cerebral ischemia and metabolic stress. Proc Natl Acad Sci U S A 98:4044-4049.

Smith T, Groom A, Zhu B, Turski L (2000) Autoimmune encephalomyelitis ameliorated by AMPA antagonists. Nat Med 6:62-66.

Sommer N, Loschmann PA, Northoff GH, Weller M, Steinbrecher A, Steinbach JP, Lichtenfels R, Meyermann R, Riethmuller A, Fontana A, et al. (1995) The antidepressant rolipram suppresses cytokine production and prevents autoimmune encephalomyelitis. Nat Med 1:244-248.

Soriano SG, Amaravadi LS, Wang YF, Zhou H, Yu GX, Tonra JR, Fairchild-Huntress V, Fang Q, Dunmore JH, Huszar D, Pan Y (2002) Mice deficient in fractalkine are less susceptible to cerebral ischemia-reperfusion injury. J Neuroimmunol 125:59-65.

Spengler RN, Spengler ML, Lincoln P, Remick DG, Strieter RM, Kunkel SL (1989) Dynamics of dibutyryl cyclic AMP- and prostaglandin E2-mediated suppression of lipopolysaccharide-induced tumor necrosis factor alpha gene expression. Infect Immun 57:2837-2841.

Sperschneider H, Neumann K, Ruffert K, Stein G (1996) Influence of recombinant human erythropoietin therapy on in vivo chemotaxis and in vitro phagocytosis of polymorphonuclear cells of hemodialysis patients. Blood Purif 14:157-164.

Spinas GA, Bloesch D, Keller U, Zimmerli W, Cammisuli S (1991) Pretreatment with ibuprofen augments circulating tumor necrosis factor-alpha, interleukin-6, and elastase during acute endotoxinemia. J Infect Dis 163:89-95.

Squadrito F, Altavilla D, Squadrito G, Campo GM, Arlotta M, Quartarone C, Saitta A, Caputi AP (1999) Recombinant human erythropoietin inhibits iNOS activity and reverts vascular dysfunction in splanchnic artery occlusion shock. Br J Pharmacol 127:482-488.

Sun Y, Jin K, Xie L, Childs J, Mao XO, Logvinova A, Greenberg DA (2003) VEGF-induced neuroprotection, neurogenesis, and angiogenesis after focal cerebral ischemia. J Clin Invest 111:1843-1851.

Villa P, Bigini P, Mennini T, Agnello D, Laragione T, Cagnotto A, Viviani B, Marinovich M, Cerami A, Coleman TR, Brines M, Ghezzi P (2003) Erythropoietin selectively attenuates cytokine production and inflammation in cerebral ischemia by targeting neuronal apoptosis. J Exp Med 198:971-975.

Viviani B, Corsini E, Galli CL, Marinovich M (1998) Glia increase degeneration of hippocampal neurons through release of tumor necrosis factor-alpha. Toxicol Appl Pharmacol 150:271-276.

Viviani B, Corsini E, Galli CL, Padovani A, Ciusani E, Marinovich M (2000) Dying neural cells activate glia through the release of a protease product. Glia 32:84-90.

Wang L, Zhang Z, Wang Y, Zhang R, Chopp M (2004) Treatment of stroke with erythropoietin enhances neurogenesis and angiogenesis and improves neurological function in rats. Stroke 35:1732-1737.

White CA, McCombe PA, Pender MP (1998) The roles of Fas, Fas ligand and Bcl-2 in T cell apoptosis in the central nervous system in experimental autoimmune encephalomyelitis. J Neuroimmunol 82:47-55.

Chapter 13

DEVELOPMENT OF NON-ERYTHROPOIETIC ERYTHROPOIETIN VARIANTS FOR NEUROPROTECTION

Lars Torup and Marcel Leist

H. Lundbeck A/S, 9 Ottiliavej, DK-2500 Valby, Copenhagen, Denmark

Abstract: Erythropoietin is well known to possess erythropoietic activity, but also tissue protection. Here, we present examples on how to separate these two activities. One possibility is to generate a short-lived variant by removal of EPO's sialic acid residues. Although asialo-EPO has a high affinity for the classical EPO-R, it lacks hematopoetic activity in vivo upon bolus injection, because of its short plasma half-life. Another approach we followed was to generate carbamylated EPO (CEPO), which has no affinity for the EPO receptor, but the same neuroprotective potency as EPO. Our data suggest that the cytoprotective signal transduction of EPO is distinct from the hematopoietic signaling machinery.

Key words: Asialo-EPO, carbamylation, chemical modification, mutation, neuroprotection,

1. INTRODUCTION

Erythropoietin is a 30-35kDalton glycoprotein containing about 40% carbohydrates. Blood-borne EPO is primarily produced in the kidney and controls the proliferation and differentiation of red blood cells in the bone marrow. EPO induces hematopoiesis by binding to the classical homodimer EPO receptor. EPO has two independent binding sites for the EPO receptor - one high affinity (Kd in nM range) and one low affinity (Kd in μM range) (Philo et al., 1996). The regulation of erythropoiesis by EPO is well established. However, the biological function of EPO is not limited to hematopoiesis. EPO and EPO-R are also expressed in other tissues, such as the brain. Through the 1990's evidence accumulated that in addition to stimulating red blood cell production EPO was able to protect neurons from dying by inhibiting apoptosis and providing trophic support. In 2000, Brines et al. reported that EPO, administrated peripherally, crossed the blood brain barrier under non-pathological conditions (Brines et al., 2000). A single intraperitoneal dose of 5000 U/kg was sufficient to ameliorate disease progression in different pathological animal models.

EPO is generally considered a safe and well-tolerated treatment of anemia. In terms of using EPO as a tissue protectant agent in patients without anemia, the hematological effects of EPO are thus to be considered as side effects. In addition to the stimulation of erythropoiesis, EPO has been shown to affect the formation and function of platelets. In line with this, EPO receptors have been detected on rat and mouse megakaryocytes (Fraser et al., 1989). Wolf et al. showed that EPO in dogs stimulated the generation of platelets (thrombocytes) and that these thrombocytes as well as the whole population of platelets were functionally hyperreactive (Wolf et al., 1997b). In a second study, using a dog model of arterio-venous shunting the group reported that EPO was pro-thrombotic (Wolf et al., 1997a). In a clinical trial on healthy volunteers, EPO increased the platelet and endothelial activation as measured by P-selectin and cE-selectin (Stohlawetz et al., 2000). Van Geet found that EPO treatment in hemodialysis patients increased the number and function of platelets which is in line with the findings by Cases that EPO improved platelet function in uremic patients (van Geet et al., 1989; Cases et al., 1992). Beguin concluded that EPO treatment induced thrombocytosis at relatively moderate doses (Beguin, 1999). These studies suggest that EPO therapy is likely to increase the risk of thrombotic event even when EPO is administered for a short period. Therefore, it would be desirable to modify EPO to obtain a molecule devoid of erythropoietic effect, but with retained tissue protective effect. In the following sections, we will give examples of different ways to dissociate the tissue protective from the erythropoietic effect by modifying recombinant human EPO.

2. ASIALO-EPO

EPO contains three carbohydrate groups, which are N-linked, and one O-linked oligosaccaride. The galactose residues of EPO's sugar part are masked by up to 14 sialic acid residues, which are the terminal sugars of the branched carbohydrate chains. In the circulation, these sialic acids are slowly removed by different sialidases as EPO "ages", and galactose residues become the new terminal residues. This allows for binding of desialylated

EPO to galactose receptors (= asialoglycoprotein receptors) in the liver, followed by internalization and digestion by lysosomes. Complete enzymatic desialylation of EPO yields asialo-EPO, a molecule with very high clearance from the circulation with a plasma elimination half-life of 1.4 min (rat, 44ug/kg IV) (Erbayraktar et al., 2003). Since the protein backbone of EPO is not affected by the sialic acid content, desialylation does not affect the interaction of EPO/asialoEPO with receptors. However, a continuous stimulation of the EPO receptor by a ligand is mandatory for stimulation of erythropoiesis. Because of the very high elimination rate, asialo-EPO is functionally without erythropoietic effect in vivo. For neuroprotection, obviously, a short trigger of the signal transduction pathway is sufficient and therefore the neuroprotective properties are retained both in vitro and in vivo.

Table 13-1 shows in quantitative terms that the affinity of asialo-EPO to the soluble EPO receptor is unchanged as well as the hematopoietic bioactivity as seen in the EPO dependent human leukemia cell line UT-7. The in vitro neuroprotective effect of asialo-EPO is similar to EPO's as seen in P19 cells subjected to serum withdrawal and in hippocampal cell cultures subjected to NMDA toxicity.

Although the plasma half-life of asialo-EPO is too short to stimulate erythropoiesis in vivo, the neuroprotective effects are retained in various animal models of disease. Asialo-EPO offered protection to the same extent as EPO in a rat model of spinal cord injury {Erbayraktar et al., 2003}. Rats were injured by transient compression of the spinal cord and dosed with EPO, asialo-EPO or vehicle (10ug/kg IV) for the first three days followed by bi-weekly dosing. Both EPO and asialo-EPO treatment significantly improved the motor scores compared to vehicle treatment. The extensive damage and edema throughout the spinal cord was reduced to a core injury in the EPO and asialo-EPO treated groups.

In a rat model of stroke (permanent middle cerebral artery occlusion) a single dose of 5-50 μg/kg IV at 90 minutes post-injury reduced the infarct size by 50% measured at 24 h post-injury.

Despite the short half-life in plasma, asialo-EPO reached the CSF at concentrations relevant for the in vitro neuroprotection (0.5-30 pM). As an alternative way to monitor the brain penetration, asialo-EPO was radiolabeled and administered peripherally. Subsequent autoradiographic analysis revealed a specific neuronal localization similar to the pattern of radiolabeled EPO. Until a few years ago it was generally believed that large glycosylated proteins like EPO were unable to cross the blood-brain barrier. However, several groups have shown that peripherally administered EPO is in fact able to enter the brain (Brines et al., 2000; Ehrenreich et al., 2004; Grasso et al., 2002; Jumbe-Nelson, 2002; Jumbe-Nelson, 2002). The presence of EPO-R on capillary endothelial cells might suggest a specific EPO transport mechanism. Recently, Martinez-Estrada and co-workers characterized such a specific, saturable mechanism responsible for EPO's transport across the blood-brain barrier (Martínez-Estrada et al., 2003).

Table 13-1. Comparison of different EPO variants

Modification	EPO-R IC50 (pM)	UT-7 EC50 (pM)	Hippocampal neurons (% protection)	P19 (% protection)	Predominant plasma half-life (H)
Rh-EPO	10	10-30	78 +/- 13	49 +/- 12	5.6
Asialo-EPO	14	10-30	71 +/- 15	45 +/- 15	0.023
CEPO	>10,000	>10,000	70 +/- 9	49 +/- 10	3.3
S100E-EPO	>10.000	>10.000	66 +/- 9	55 +/- 15	n.d.

3. NEUROPROTECTION VS HEMATOPOIESIS

A number of recent findings point to the possibility that action of EPO on hematopoiesis and cytoprotection may involve different signal transduction cascades (e.g. Digicaylioglu, Lipton, 2001). Masuda et al. reported that cultured rat astrocytes produced EPO and that the production was dependent on the oxygen tension. He proposed that EPO produced by astrocytes acted on the EPO receptor on neurons in a paracrine fashion (Masuda et al., 1994). Neurons indeed express EPO-R and EPO's affinity to the neural receptor appears to be lower (10-20 nM) than the erythroid receptor affinity (100-200 pM). Sasaki and coworkers found that the EPO receptor expressed on PC12 and SN6 cells which both have neuronal characteristics has a correspondingly lowered affinity (Masuda et al., 1993). This discrepancy in receptor affinity between peripheral and neuronal tissue suggests that the neuronal EPO receptor might be different from the EPO receptor found in erythroid cells. Further experiments revealed different sizes of receptor associated proteins and receptor molar weight dependent on the cell type. The classical EPO receptor is a homodimer, but that does not exclude that heterodimers or oligomers could constitute functional EPO receptors in non-erythroid cells. It has for instance been reported that EPO can form functional receptor complexes with other cytokine receptors such as the CD131 receptor (Beta common chain) (Hanazono et al., 1995; Jubinsky et al., 1997).

The finding that EPO has a neurotrophic sequence unrelated to the binding sites to the classical EPO receptor and that this 17-mer peptide sequence is non-hematopoietic further suggests that cytoprotection and hematopoiesis are mediated via two distinct mechanisms (Campana et al., 1998).

Ideally, one would want an EPO molecule not just without functional lack of erythropoiesis in vivo, but a molecule that does not interact at all with the bone marrow EPO receptor. From the literature, it is known that a small chemical modification or a single mutation is enough to abolish the affinity to the EPO receptor (Grodberg et al., 1996). We speculated that it would be possible to engineer the EPO molecule to loose the affinity to the bone marrow EPO receptor, while retaining the cytoprotection.

4. CARBAMYLATED EPO (CEPO)

There are many possibilities to chemically engineer the EPO molecule. The lysines, arginines, tyrosines and the carboxyl groups on the EPO molecule can be chemically altered e.g. by carbamylation, amidation, trinitrophenylation or acylation. We explored in particular the possibility to carbamylate the seven lysines in the EPO molecule to generate homocitrulline residues (Leist et al., 2004). Under normal physiological conditions 0.8% of urea is converted to cyanate in human plasma. This readily reacts with EPO (Satake et al., 1990) to form partially carbamylated EPO (Mun, Golper, 2000). Partially carbamylated EPO is thus a naturally occurring form of EPO in man. Surprisingly, we found that although the fully carbamylated EPO molecule we produced (CEPO) is devoid of interaction with the classical EPO receptor the cytoprotective properties were retained (Leist et al., 2004). CEPO has at least 1000 fold less affinity to the soluble EPO receptor and likewise more than 1000 fold lower effect in stimulation of the UT-7 erythropoietic cells compared to EPO (Fig. 13-1). In addition, CEPO does not induce the JAK/STAT signaling seen with EPO in BaF/3 cells transfected with the EPO receptor. Moreover, CEPO is not capable of antagonizing EPO in the UT-7 cell even at concentrations 300 times higher than EPO. Despite the clear loss of EPO receptor interaction CEPO is as efficacious as EPO in hippocampal cell cultures subjected to NMDA. In P19 cells challenged by serum withdrawal CEPO is as efficacious as EPO in rescuing the cells (table 13-1).

Obviously, it is of major importance that an engineered EPO variant enters the CNS to at least the same extend as EPO. Indeed, CSF levels of CEPO after intravenous dosing (44ug/kg CEPO) were comparable to the levels of EPO. Even 24 h after intravenous dosing, the CSF levels were significantly above baseline. As opposed to the asialo form of EPO the carbamylation does not alter the kinetic profile substantially. The predominant plasma half-life of CEPO in rats is in the same range as that of EPO (3-6 h). After subcutaneous injection the plasma concentrations was above 2 nM (the erythropoietic threshold for EPO) for more than 20 h. Thus

CEPO circulates long enough to stimulate erythropoiesis. To rule out the possibility that CEPO was stimulating erythropoiesis through a non-EPO receptor mechanism, several experiments were conducted to examine the effects of CEPO on hemoglobin and hematocrit levels. Bi-weekly subcutaneous dosing with up to 500 μg/kg CEPO did not increase the hematocrit in mice over 10 weeks. In another experiment mice were subcutaneous dosed daily for 4 weeks. The mice dosed from 10 to 200 μg/kg CEPO had no increase in hemoglobin or hematocrit as opposed to a group of mice receiving 10 μg/kg EPO, which had a significant increase in hematocrit. It was further tested whether CEPO could antagonize the erythropoietic effect of EPO in vivo. Even ten-fold higher dose of CEPO (50μg/kg/day) could not antagonize the effect of EPO (5 μg/kg/day) on hematocrit.

Table 13-2. Effect of EPO variants in animal models

	EPO	Asialo-EPO	CEPO
Hematopoiesis	+	÷	÷
Stroke	+	+	+
Spinal cord injury	+	+	+
EAE	+	+	+

+ = biological activity; ÷ = lack of biological activity

The data from the in vitro cell death models suggest that CEPO is at least as efficacious and potent as EPO in various animal disease models of acute and chronic neurodegeneration. This was tested in detail. In a rat model of stroke, CEPO decreased the infarct volume by 50% in a dose-range from 5-50μg/kg IV when measured 24 h after injury compared to a saline treated group. The tissue protection was retained even when the dosing was postponed to 4 h after onset of the injury. A similar broad time window of opportunity has been reported for EPO (Brines et al., 2000). The tissue protection correlated well with an ipsilateral reduction in inflammatory markers such as Interleukin-6 and Monocyte chemoattractant protein-1.

In a rat model of spinal cord injury the neurological function was examined over 42 days during chronic treatment with EPO, CEPO or saline (Leist et al., 2004). The CEPO treated group recovered fully from the injury with a neurological score significantly higher than the saline or the EPO-treated group. Delaying the CEPO treatment for 24hrs was as efficacious as when the treatment is initiated immediately after the injury. Even a delay of 72hrs significantly improved the recovery compared to saline treated rats.

Furthermore, CEPO was tested in a mouse model of experimental autoimmune encephalomyelitis (EAE) (Leist et al., 2004). CEPO treatment 3 times per week significantly improved the neurological outcome in the mice immunized with myelin oligodendrocyte glycoprotein (MOG) to induce the

EAE. Improvement in functional outcome was seen even when CEPO treatment was commenced 4 weeks after the mice had reached a plateau in symptoms.

Lastly, CEPO was tested in a model of diabetic neuropathy induced by streptozotocin. CEPO treatment (3 times per week subcutaneously) improved the nerve dysfunction as assessed by the thermal nociceptive threshold in the hotplate test.

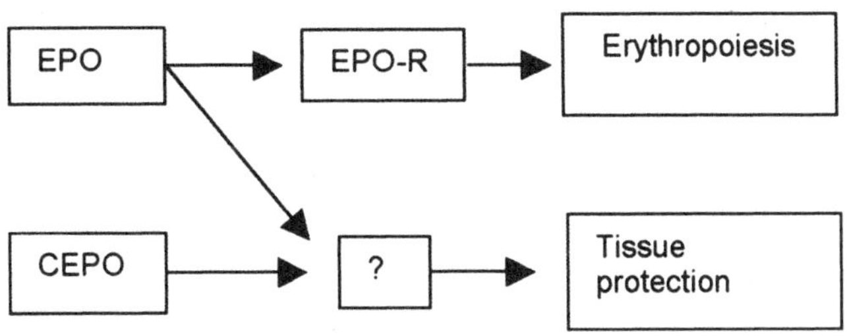

Figure 13-1. Biological activities of EPO and CEPO

5. OUTLOOK

In this chapter we have given examples of different ways to modify EPO to dissociate the erythropoietic effect from the tissue protective effect. While the principle behind asialo-EPO is dissociation based on altering the pharmacokinetic profile, the different effects of CEPO are based on altered pharmacodynamics (no interaction with the EPO receptor). Engineered proteins in clinical use differ from the natural template by altered stability, kinetic profile or antigenicity. In contrast to this, CEPO is an example of a new class of compounds with a new bioactivity profile. It has been designed to have a similar pharmacokinetic profile as EPO, but exhibiting a new mode of action at the molecular level. This model of action can best be explained by interaction with an alternative receptor transducing tissue protection signals (see Fig. 13-1). This receptor is unlikely to be the classical EPO receptor homodimer. One possibility is that CEPO binds a heterodimer involving only one EPO receptor. Such a receptor would constitute a new pharmacological target, and the availability of CEPO as selective ligand will allow pharmacological characterization and exploration of the underlying biology.

REFERENCES

Beguin Y (1999) Erythropoietin and platelet production. Haematologica 84:541-547.

Brines ML, Ghezzi P, Keenan S, Agnello D, de Lanerolle NC, Cerami C, Itri LM, Cerami A (2000) Erythropoietin crosses the blood-brain barrier to protect against experimental brain injury. Proceedings of the National Academy of Sciences of the United States of America 97:10526-10531.

Campana WM, Misasi R, Brien JS (1998) Identification of a neurotrophic sequence in erythropoietin. International journal of molecular medicine 1:235-241.

Cases A, Escolar G, Reverter JC, Ordinas A, Lopez PJ, Revert L, Castillo R (1992) Recombinant human erythropoietin treatment improves platelet function in uremic patients. Kidney international 42:668-672.

Digicaylioglu M, Lipton SA (2001) Erythropoietin-mediated neuroprotection involves cross-talk between Jak2 and NF-kappaB signalling cascades. Nature 412:641-647.

Ehrenreich H, Degner D, Meller J, Brines M, Béhé M, Hasselblatt M, Woldt H, Falkai P, Knerlich F, Jacob S, von Ahsen N, Maier W, Brück W, Rüther E, Cerami A, Becker W, Sirén AL (2004) Erythropoietin: a candidate compound for neuroprotection in schizophrenia. Molecular psychiatry 9:42-54.

Erbayraktar S, Grasso G, Sfacteria A, Xie Q, Coleman T, Kreilgaard M, Torup L, Sager T, Erbayraktar Z, Gokmen N, Yilmaz O, Ghezzi P, Villa P, Fratelli M, Casagrande S, Leist M, Helboe L, Gerwein J, Christensen S, Geist MA, Pedersen LØ, Cerami HC, Wuerth JP, Cerami- Anthony, Brines M (2003) Asialoerythropoietin is a nonerythropoietic cytokine with broad neuroprotective activity in vivo. Proceedings of the National Academy of Sciences of the United States of America 100:6741-6746.

Fraser JK, Tan AS, Lin FK, Berridge M, V (1989) Expression of specific high-affinity binding sites for erythropoietin on rat and mouse megakaryocytes. Experimental hematology 17:10-16.

Grasso G, Buemi M, Alafaci C, Sfacteria- Alessandra, Passalacqua M, Sturiale A, Calapai-Gioacchino, De Vico G, Piedimonte G, Salpietro F, Tomasello F (2002) Beneficial effects of systemic administration of recombinant human erythropoietin in rabbits subjected to subarachnoid hemorrhage. Proceedings of the National Academy of Sciences of the United States of America 99:5627-5631.

Grodberg J, Davis KL, Sytkowski AJ (1996) Functional and structural role of arginine 103 in human erythropoietin. Archives of biochemistry and biophysics 333:427-431.

Hanazono Y, Sasaki K, Nitta H, Yazaki Y, Hirai H (1995) Erythropoietin induces tyrosine phosphorylation of the beta chain of the GM-CSF receptor. Biochemical and biophysical research communications @208:1060-1066.

Jubinsky PT, Krijanovski O, I, Nathan DG, Tavernier J, Sieff CA (1997) The beta chain of the interleukin-3 receptor functionally associates with the erythropoietin receptor. Blood 90:1867-1873.

Jumbe-Nelson LN (2002) Erythropoietic agents as neurotherapeutic agents: what barriers exist? Oncology 16:91-107.

Leist M, Ghezzi P, Grasso G, Bianchi R, Villa- Pia, Fratelli M, Savino C, Bianchi M, Nielsen-Jacob, Gerwien J, Kallunki P, Larsen AK, Helboe- Lone, Christensen S, Pedersen LO, Nielsen M, Torup L, Sager T, Sfacteria A, Erbayraktar S, Erbayraktar- Zubeyde, Gokmen N, Yilmaz O, Cerami HC, Xie QW, Coleman T, Cerami A, Brines M (2004) Derivatives of erythropoietin that are tissue protective but not erythropoietic. Science 305:239-242.

Martínez-Estrada OM, Rodríguez ME, De Vicente E, Reina M, Vilaró S, Fabre M (2003) Erythropoietin protects the in vitro blood-brain barrier against VEGF- induced permeability. The European journal of neuroscience 18:2538-2544.

Masuda S, Nagao M, Takahata K, Konishi Y, Gallyas F, Jr., Tabira T, Sasaki R (1993) Functional erythropoietin receptor of the cells with neural characteristics. Comparison with receptor properties of erythroid cells. The Journal of biological chemistry 268:11208-11216.

Masuda S, Okano M, Yamagishi K, Nagao M, Ueda M, Sasaki R (1994) A novel site of erythropoietin production. Oxygen-dependent production in cultured rat astrocytes. The Journal of biological chemistry 269:19488-19493.

Mun KC, Golper TA (2000) Impaired biological activity of erythropoietin by cyanate carbamylation. Blood purification 18:13-17.

Philo JS, Aoki KH, Arakawa T, Narhi LO, Wen J (1996) Dimerization of the extracellular domain of the erythropoietin (EPO) receptor by EPO: one high-affinity and one low-affinity interaction. Biochemistry 35:1681-1691.

Satake R, Kozutsumi H, Takeuchi M, Asano K (1990) Chemical modification of erythropoietin: an increase in in vitro activity by guanidination. Biochimica et biophysica acta 1038:125-129.

Stohlawetz PJ, Dzirlo L, Hergovich N, Lackner E, Mensik C, Eichler H, Kabrna E, Geissler K, Jilma B (2000) Effects of erythropoietin on platelet reactivity and thrombopoiesis in humans. Blood 95:2983-2989.

van Geet C, Hauglustaine D, Verresen L, Vanrusselt M, Vermylen J (1989) Haemostatic effects of recombinant human erythropoietin in chronic haemodialysis patients. Thrombosis and haemostasis 61:117-121.

Wolf RF, Gilmore LS, Friese P, Downs T, Burstein SA, Dale GL (1997a) Erythropoietin potentiates thrombus development in a canine arterio- venous shunt model. Thrombosis and haemostasis 77:1020-1024.

Wolf RF, Peng J, Friese P, Gilmore LS, Burstein SA, Dale GL (1997b) Erythropoietin administration increases production and reactivity of platelets in dogs. Thrombosis and haemostasis 78:1505-1509.

INDEX